CW01459806

# MOVING PEOPLE

*Charles Seeberger invented the word Escalator, Otis patented it, and the incredibly efficient passenger carrying machine was born.*

*First published 2000*

*ISBN 1-886536-25-2*

*Published by Elevator World, Inc.*
*P.O. Box 6507, 356 Morgan Avenue*
*Mobile, Alabama 36606 U.S.A.*

*Printed in the U.S.A.*
*by Davidson Printing Co.*

*Designed by Grey Lipley and John Gale, London.*

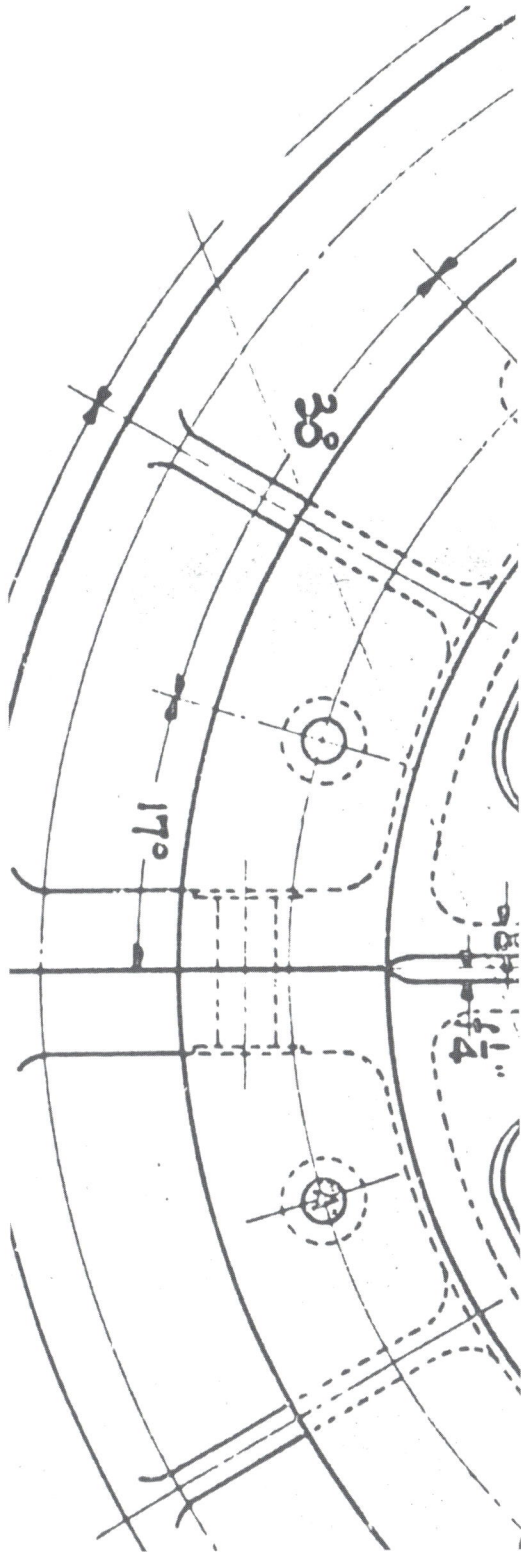

# MOVING PEOPLE
## From Street To Platform

100 years **UNDERGROUND**

# ESCALATORS
## People Moving Machinery

**Author**
Ray Orton

**Researcher**
Nick Gaw

**Book Design**
Grey Lipley
John Gale

## ACKNOWLEDGMENTS

Writing and editing a book is pleasurable yet hard work. It takes time and patience and often relies upon the help and assistance of other people. To all these other people who have helped me with my research, I would like to take this opportunity to offer my sincere thanks.

To Nick Gaw for his help, support and advice throughout the publication, to John Fryer, Matthew Self and Norman Cohen for getting me started and to Mike Ashworth and David Ellis of the Covent Garden Transport Museum.

My special thanks to John Gale and Grey Lipley for their enthusiastic and creative input, without which this book would not have been possible.

I would also like to thank Dennis Burrows and Ken Holm of the Otis Elevator Company Archives, David Cooper of CIBSE. The London Otis staff, with especial thanks to Dave McGraw and Mike Hirst and to Ron Wanless and Jeff Midgley of O&K and Nick Chappelle of CNIM.

Finally, thanks to Mike Horne of LUL and Paul Hadley, who both contributed to the final information on the Spiral Walkway at Holloway Road.

Ray Orton.

## CHAPTERS

## INTRODUCTION

Escalators have been in public service on the London Underground since the first installation of the two Seeberger machines at Earls Court Station in 1911. The public acceptance of these machines made them the most convenient form of vertical passenger transportation, within London's Underground Railway system.

This book traces the history and development of these machines, from Jesse Reno's Spiral in 1906, to the most modern installations of the Jubilee Line Extension, currently undergoing construction. My personal, close affinity with these machines and the people who manufacture, maintain and install them, has led to the writing of this book. I have worked for London Underground for over twenty-five years and have never lost my initial enthusiasm and interest in these unique machines and their environment.

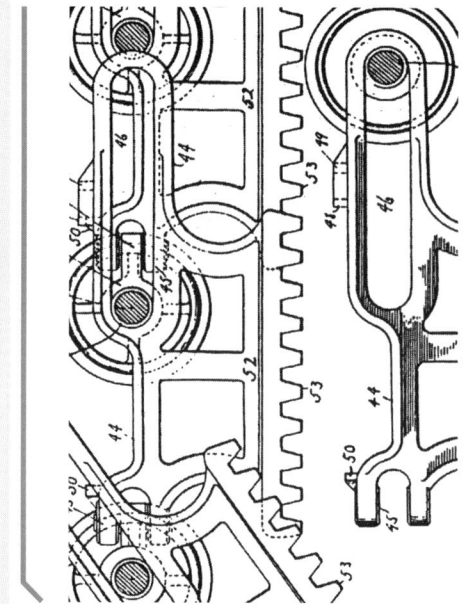

*Alfred Newton* filed, and was granted, provisional protection of his invention for moving stairs in England in 1858. He did not submit a final specification, leaving the way clear for Nathan Ames to claim credit for the first patent.

6

I hope that after reading the book, some of my enthusiasm rubs off, and members of the travelling public, can understand and appreciate the importance of the escalator to the Underground network.

Ray Orton.

# THE EARLY UNDERGROUND RAILWAY

## CHRONOLOGICAL HISTORY
## ( TIME LINE )

## THE EARLY UNDERGROUND RAILWAY

The world's first underground railway was opened on the 10th January 1863 and ran from Paddington to Farringdon via Baker Street. The carriages were pulled by specially designed steam locomotives that did not expel steam in the tunnels but were very noisy when relieving steam pressure at stations. The passengers, undertaking their initial journey, experienced the thrilling but eerily gloomy conditions of the tunnels, yet still were not overawed. The railway was an immediate success.

This early success and ability to transport passengers relatively quickly, beneath London's very crowded streets, led to construction of other, very fragmented and privately owned railway lines. This uncoordinated expansion of the system carried on throughout the late 1800's, until the arrival of Charles Tyson Yerkes. Yerkes was an American entrepreneur who arrived in London in 1900 and by a succession of skillful business manoeuvres gained control and linked several of the privately owned railways. This was the beginnings of the network we now know as London Underground.

Left: Queens Park Station, 1914.
Above: Trafalgar Square Station, 1925.
Right: Kilburn Park, vertical shaft, 1914.

Yerkes moved to Chicago in the 1880's and by 1886 had gained control of the North, then West division Tramways. He made many improvements to the system and created a huge suburban network, including the elevated railway. In 1889, he sold out and moved to New York.

By 1900, he had moved to London and soon gained control of the Charing Cross, Euston and Hampstead Line which he immediately planned to extend to Golders Green.

By March 1901, he had purchased the District Line, closely followed by the Metropolitan and the Brompton and Piccadilly Circus Lines. Next came the Great Northern and Strand Scheme, which linked with Piccadilly Circus and Holborn and then to Earls Court where there was a link to Finsbury Park. The combined routes were known as The Great Northern, Piccadilly and Brompton Railway. In 1902, Yerkes took over the Baker Street and Waterloo Line; he registered the new company as the Underground Electric Railways Company of London Limited.

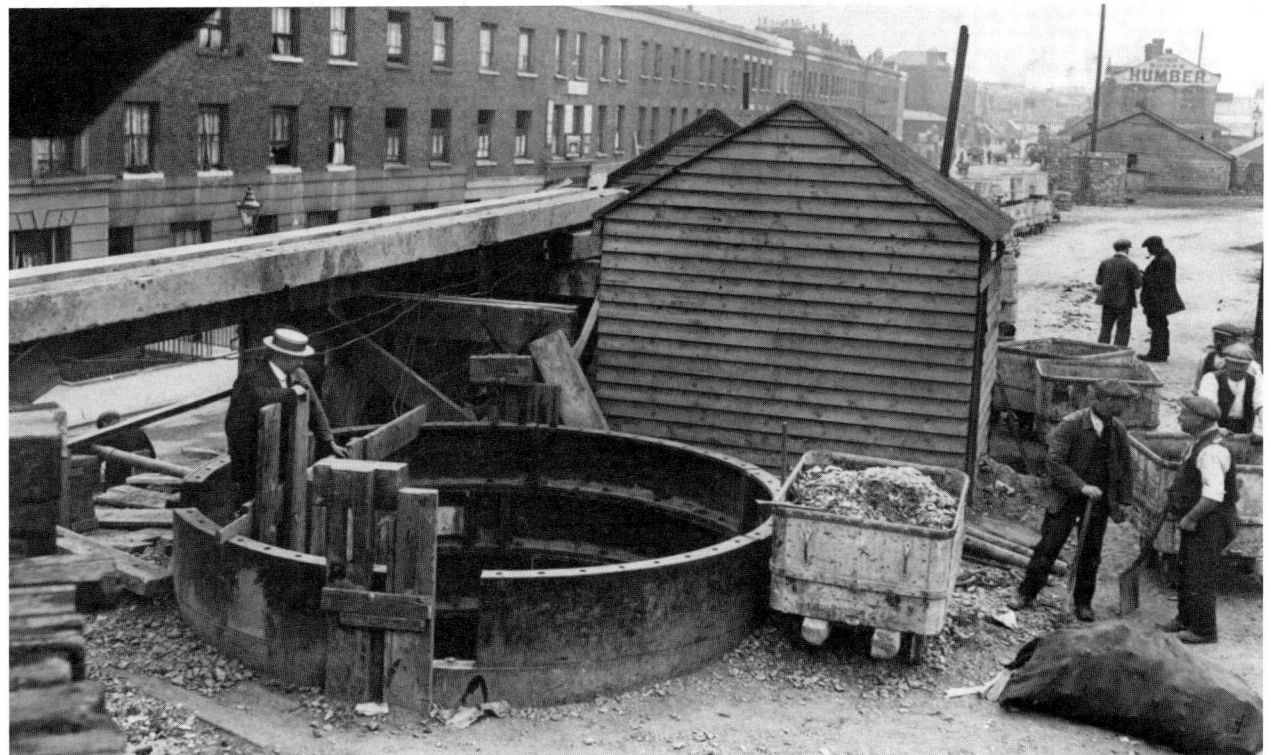

*Charles Tyson Yerkes* was born in Philadelphia USA in 1837. At the age of 22, he opened his own brokerage office, quickly followed by the purchase of his own banking house. Forced into bankruptcy, he was indicted for embezzlement and sentenced to 33 months in the penitentiary. He was pardoned after seven months and released.

His network was now beginning to take shape and Yerkes again began to expand the railway. In 1907, 16,000,000 passengers per year were using the system and London began to benefit from this magnificent Underground Railway.

As the Underground Railway expanded to encompass all of the Greater London Boroughs, it was also forced to go deeper underground. A combination of both the older cut and cover method for near surface stations, and the Greathead Shield or Price Rotary Excavator, were developed to increase the efficiency of advanced deep level tunnel boring.

Deep tunnels serving street level stations dictated that passengers would have to be transported from platform level up to street level. For this purpose and because of restrictions imposed by tunnelling in a major city, high-rise escalators were developed.

Prior to 1911, all the vertical transportation of passengers had been undertaken by station lifts. The two Seeberger escalators installed at Earls Court Station were so successful that lift replacement was gradually introduced. The escalator became the prime mode of passenger transportation from station platform to street level.

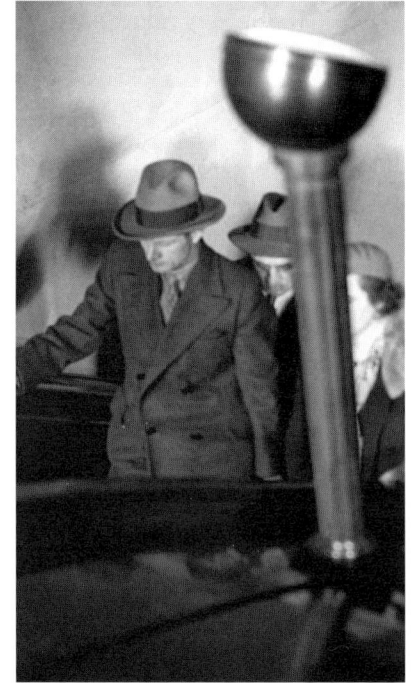

As the fragmented parts of the railway joined up or were extended, station designs of the period were unable to cope. London had never before witnessed a development on this scale or an engineering project so vast. The managers of the time had no knowledge of passenger numbers, drainage or station ventilation, let alone passenger flow or passenger control. New and innovative solutions would have to be found quickly if the railway was to function.

*Left: Cutting clay at the work surface.*
*Above: Highgate Station, 1931.*

As demand on the system increased, it was expanded to accommodate this new-found popularity. By the 1930's, the system had become an accepted part of London's Public transport. The popularity and efficiency of the Tube greatly enhanced London's status as a capital city.

New escalator designs were introduced and the thirties saw the inauguration of the M series machines. These escalators were designed specifically for vertical rises in excess of 90 feet and were to become the Rolls Royce of the escalator world. The marvellous design of these machines set the standard for the years ahead and ensured that the demands of the Underground Railway could be met.

These escalators ran at speeds of up to 180 ft/min, 20 hours per day, seven days per week, three hundred and sixty-five days per year. This ensured that the Underground Railway offered an incomparable service to the travelling public.

*Above: Oval Station escalator controller.*
*Right: Booking Hall, Piccadilly Circus Station, 1928.*

The design standard of these machines is so high that even after sixty years, 14 MH and 22 MY machines are still in passenger service. Since installation, they have withstood the rigours of the general public and bombing by the Luftwaffe during World War II. Like our present day road system, who would have guessed, in 1920, that there would be so much demand put on the Underground Railway or that it would become so popular.

## CHRONOLOGICAL HISTORY TIME LINE

Reno
Inclined
Escalator

On 9th August 1859, United States Patent No. 25076 was granted to Nathan Ames for a revolving stairs. These revolving stairs were designed to move in the form of an equilateral triangle. Passengers would board at the bottom, travel up to the apex of the triangle, then hop off sideways. To descend, a reversal of this process was necessary but the great agility and coordination required by the passenger would have made this a very dangerous practice.

As far as is known, these moving stairs were never built and remained only an idea. Nathan Ames was not heard of again but his ancient, unused patent was the first recorded patent for revolving stairs. No more patents for moving stairs were granted until nearly the turn of the century when, within a few years, 3 designs for moving stairways were submitted and 3 patents granted.

1891 Jesse Reno.    1892 George H Wheeler.    1898 Charles D Seeberger.

In September 1895, the first Reno moving stairway was installed as a pleasure ride at Coney Island, Brooklyn, New York. It had a vertical rise of 7 feet, was inclined at 25 degrees with a step width of 20 inches and running speed of about 75 feet per minute.

George H. Wheeler and Charles D. Seeberger, joined forces and came up with a greatly improved design. Seeberger invented the word escalator to describe this machine.

**1899**
Seeberger sold the design to the Otis Elevator Company.
Competition between the two designs, Jesse Reno and Otis, for sales of their respective machines was hotting up. The Reno Electric Stairway and Conveyors Organisation had been selling and exhibiting their machines throughout the USA and Europe.

**1899**
A Reno machine was installed at London's Crystal Palace. Passengers had to pay one old penny for a ride and W.P. Dempsey wrote a music hall song about the machine called Up The Sliding Stairs.

**1900**
At the Paris Exposition, a Reno, moving pallet-type walkway, was installed.

**1901**
A moving stairway was fitted at Seaforth Station on the Liverpool Overhead Railway, making life a great deal easier for the 20th century traveller.

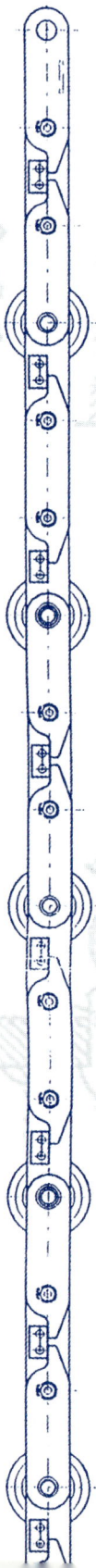

1906

1911

1911

Jesse Reno's inventive talents for the design of moving stairways reached their peak with the building, at Holloway Road Station, of a spiral moving walkway. This machine was built for the opening of the Great Northern Piccadilly and Brompton Railway and was installed in a 23 foot diameter shaft with a rise of about 35 feet.

Otis takes over the Reno Electric Stairway and Conveyor Organisation and now manufactures two types of escalator, the Seeberger or A type and the Reno or Duplex cleat type. At this time, Otis registered the word 'Escalator' as a trade name. Consequently only Otis were authorised to spell it with a capital E. This is no longer the practice due to the expiration of the patent in 1950.

*Otis A Type escalator - Earls Court 1911*

The first two escalators to be used as passenger conveyors by London Transport were installed at Earls Court Underground Station in October 1911. They were the Seeberger or A type escalator.

These first machines received a mixed reception so to allay the fears of passengers, London Transport employed a man with one leg called Bumper Harris. He rode up and down the escalators, encouraging people to board.

These two A type escalators were the forerunners of the heavy-duty escalator and between 1911 and 1915, twenty two machines were installed on the Underground system. Prototypes, however, have many peculiarities of design and the A type escalator was no exception.

13

Following the great success of the A type escalator, a programme for replacing lifts with escalators was undertaken and, between 1924-1929, sixty five LHD type escalators were installed. Piccadilly Circus Station with eleven machines was the showpiece of its day.

The LHD escalator was a reversible machine designed for vertical rises up to 60 feet. Although it outwardly resembled the A type with its wooden balustrade and shunt landings, the step-band of the LHD was driven by two step chains. Another interesting development of this machine was that the step chain wheels were an integral part of the step chain assembly.

In December 1924, No. 2 escalator at Clapham Common Station was fitted with cleated steps and combplates making it possible for passengers to move straight on and off at each landing. The new arrangement was so successful that all the shunt landing machines were then modified to accept cleated steps and combplates.

Machines with shunt landings were refurbished during the 1930's and incorporated eighteen of the A type machines resulting in them being known as MA type escalators. In 1937, as part of this programme, the first zinc alloy, cleated steps were fitted to Kilburn Park No. 2 escalator.

As the London Transport Underground system expanded, the 1930s also brought with it the arrival of a new series of escalators known as the M type machines. The MA was the modernised Seeberger. The M, MY and MH were designed for vertical rises ranging from 15 feet at Chancery Lane (M), to 90 feet at Leicester Square (MH). These escalators, like the L type, were all inclined at 30 degrees.

MA and L type escalators continued in service until 1963 when a modernisation programme commenced. The first escalator to be modernised was Oval No.1. This was the first machine fitted with closely spaced, aluminium cleated step treads, and comb sections and was known as the LHDM type escalator.

*J. & E. Hall escalator, Alperton 1955.*

*Oval No. 2 Escalator, 1963 LHD.*

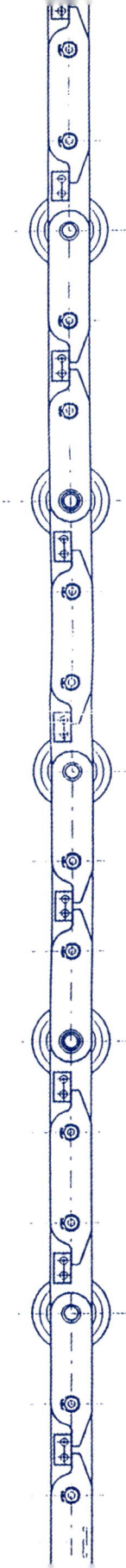

At the same time in 1963, the decision to proceed with the construction of the Victoria Line, created a requirement for another 52 escalators. In response, Otis developed the MH-A and MY-A escalators. The construction and design was basically the same as that of the MH and MY escalator with several new features.

At this time the only other escalators not built by Otis, were a J. & E. Hall machine installed at Alperton Station in 1955, and an Eggers Kehrhahn machine installed at Old Street Station. This combination of escalators gave stalwart service throughout the 1960s and 1970s, the MH and MH-A coping with the higher vertical rises, the MY, MY-A and LHDM the rises up to 60 feet.

*Eggers Kehrhahn, Old Street No. 3.*

1972

Six Otis RAC escalators, built by Ascinter in France, were installed at South Kensington Station with a further six installed at Heathrow Airport in 1987.

Toward the end of the 1970s, Otis decided to stop the manufacture of their British machines and the escalator section of their Liverpool factory closed. Escalator manufacture was transferred to their West German plant. The first German machine to be installed on the London Underground was at Charing Cross Station. The machine was prefixed HD and was of the factory assembled type. These machines are built and tested in the German factory, then broken down into sections for ease of transport and re-assembly. Slight problems were experienced during installation and testing, but the machine entered passenger service in 1980.

1980

About this time (early 1980's), the decision was taken to start pulling out the MH escalators and replace them with new machines. (In my opinion, it was a very unwise decision, I feel that, at this time, the MH escalator should have been refurbished and modernised.) However, faced with no alternative British machine available and no desire to modernise the MH, it was decided to build an HDB escalator within the existing truss-work of a high vertical rise escalator. The site chosen was Archway No. 1 and the installation commenced.

Careful cooperation between London Transport (LT) and Otis engineers ensured that the best design features of the HDB were accommodated within the MH truss-work and, after extensive modification, the machine was ready for testing. The escalator finally passed all the test procedures and entered passenger service on 13th July 1980.

**1980**

**1983**

**1984**

**1987**

**1989**

**1989**

**1992**

The rest of the 1980's and early 1990's seemed to develop into a competitive period for escalator manufacturers, all trying to produce a public service escalator for London Transport.

The first machine built by CNIM, a large French consortium, was installed in North London and entered service at Kentish Town Station.

Otis in Germany designed a brand new machine, the HDC. It was first used at Hyde Park Corner Station and closely followed by the installation of two machines at Marble Arch.

Pantin, an Essex conveyor firm, build a London Transport designed escalator and install it at Manor House Station.

Otis 'design and install' two MHB machines at Bounds Green, closely followed by Leicester Square.

APV buy out Pantin and install three machines at Kings Cross, one of London's busiest main line stations, serving the North of England.

CNIM install two machines at Kings Cross and win a prestigious contract to build the largest machines in Europe at Angel Station.

*Above: Leicester Square Station*
*Right: CNIM escalator, Angel Station, 1992.*

**1993**

O&K escalators win a contract to install fourteen machines at the Bank Station, followed by the contract for the Jubilee Line extension, one of the biggest contracts ever tendered for escalators.

**1996**

O&K take over APV and consolidate their links with London Underground, ensuring the retention of their unique experience. There follows extensive modernisation of the equipment in preparation for the opening of the Jubilee Line.

**1999**

This new underground railway is created deep beneath the streets of London to extend the capital's existing public transport infrastructure to the Millennium Dome for the Year 2000 celebrations.

UNDERGROUND

North Greenwich

*Top: Bank Station.   Above and Right: North Greenwich Station.*

| 1876 | Sioux win the Battle of Little Bighorn and kill Custer. • Alexander Bell invents the telephone. |
|---|---|
| 1880-1 | Australian bank robber Ned Kelly is executed. • Tsar Alexander 11 of Russia is assassinated. |
| 1885 | Gottlieb Daimler produces first motorcycle. • Karl Benz makes the first practical petrol car. |
| 1886 | Charles Tainter makes wax discs for Edison's phonograph. • Statue of Liberty is dedicated. |
| 1891-2 | Work begins on Trans-Siberian railway, which takes 13 years to complete. • Ellis Island opens. |
| 1895 | Wilhelm Rontgen discovers X-rays. • Lumiere brothers present the first film show in Paris. |
| 1898 | Spanish-American War, Cuba wins independence. • Pierre and Marie Curie discover radium. |
| 1900 | Kodak launch the Box Brownie camera. • British troops relieve Mafeking in the Boer War. |
| 1903-4 | Wright brothers' first powered flight. • Eugene Freyssinet creates first prestressed concrete. |
| 1907 | First facsimile transmission sent from Paris to London. • Bakelight, the first plastic, is invented. |
| 1912 | The sinking of the Titanic. • In the UK, Charles Belling develops the first effective electric fire. |

**Events List**

*Many historic events and exciting discoveries were taking place while London Underground evolved. You may find some of the things that happened during this period surprising.*

# THE FIRST ESCALATORS

## THE SPIRAL WALKWAY
## SEEBERGER MACHINES

999,885.

Patented Aug. 8, 1911.
11 SHEETS—SHEET 8.

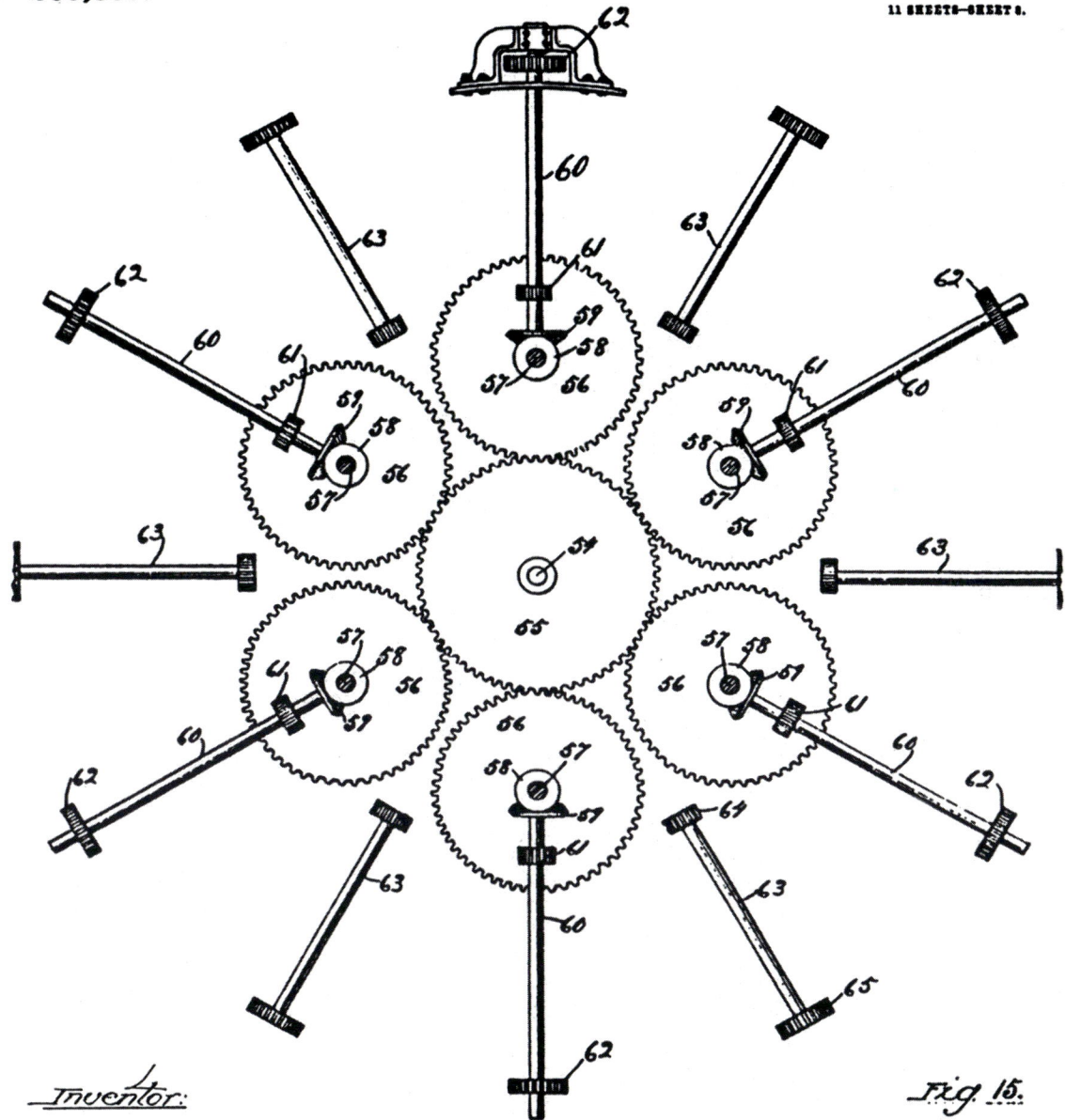

Inventor:

Fig. 15.

When Charles Tyson Yerkes' empire expanded and the number of passengers increased to more than 16,000,000 per year, it became obvious that another form of vertical passenger transport was required to complement the lifts.

In 1906, Jesse W. Reno had at his own cost, installed his uniquely magnificent Spiral Walkway at Holloway Road Station. The Spiral Escalator reputedly entered service for one day at the opening of the Great Northern, Piccadilly and Great Brompton Railway but was then removed from passenger service.

**INVENTOR RENO'S ENDLESS STAIRWAY FOR THE BRIDGE**

In 1892, **Jesse Wilford Reno** starts to build his first inclined elevators through the newly formed The Reno Inclined Elevator Company.

It was not until 1911, when Otis installed their revolutionary A or Seeberger type escalators at Earls Court Station, that the travelling public first felt the benefit of these radical machines.

Jesse Wilford Reno was born on 4th August 1861 at Fort Leavenworth, Kansas. A bright and inventive child, he had drawn up and formulated his idea for an inclined elevator at the age of sixteen. He completed his education at Lehigh University, Pennsylvania and graduated in 1883. His early career took him to Colorado and then to Americus, Georgia, where he is credited with building the first Electric Railway in the south of the United States.

Reno had submitted the first application for a patent of what he describes as a new and useful endless conveyor or elevator, on the 2nd January 1891. The application was granted and the patent became effective on the 15th March 1892. The machine was built and installed at the Iron Pier, Coney Island, Brooklyn, as a pleasure ride, in September 1895, and is officially recognised as the first escalator.

His machines were now being sold and exhibited in Europe as well as in the USA. However, he did not just confine his inventiveness to escalators and in 1896, developed plans for the building of the New York subway. This was to be a double decked underground system which could be built within three years.

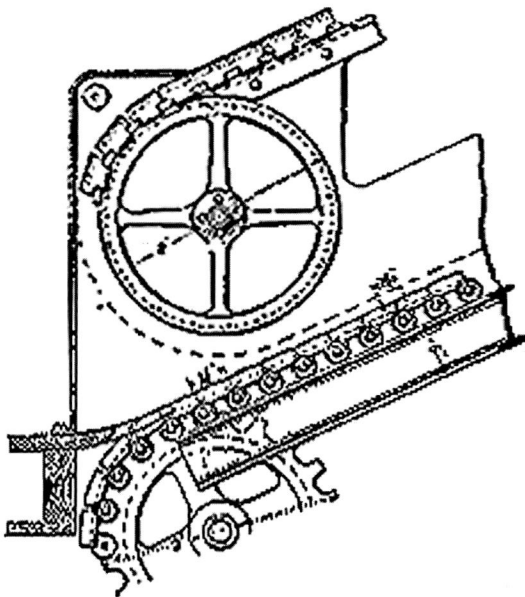

*The two illustrations on this page are examples of Reno's Early Stairway and Patented Inclined Elevator.*

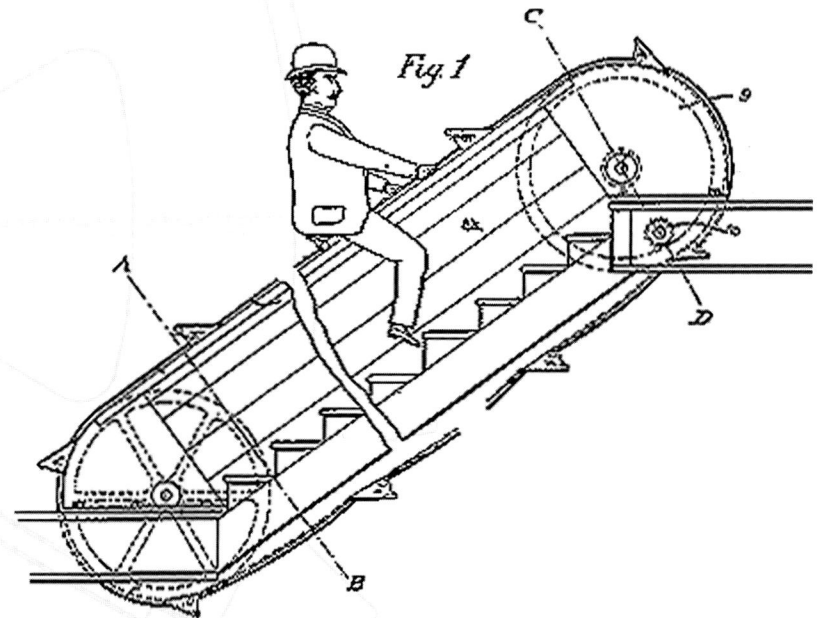

This plan was not accepted, and Reno married, travelled the world, and moved to London. Here, he opened his new company, The Reno Electric Stairways and Conveyors Ltd in 1902, with offices located at 119, Finsbury Pavement, London. During this period, he met William Henry Aston with whom he went on to build the Spiral walkway.

He must have approached Charles Tyson Yerkes for permission to build his machines, and having made the successful connection, I am very surprised that he did not become more involved with London Underground.

Jesse Reno's experience and design submissions for the New York subway must have been known to Yerkes, and his sheer expertise and inventive talent must have impressed the railway engineers of the time. His moving palette type machines were being installed throughout Europe, Great Britain and the United States. Following his move to London in 1900, he appears to have concentrated his inventive talents on the unique Spiral Walkway.

In 1902, his first efforts at Spiral Walkway design were realised. The first machine appeared at the Earls Court Exhibition. This machine appears to have run, as a fairground ride, for about four years. It was a very slow moving amusement ride which was intriguingly advertised as depicting a 'trip through the Pyrenees.'

*Right: Spiral under construction, Holloway Road, 1906.*

*Far right: Spiral stair tread, chain and guide.*

*The failure of the Spiral Escalator to enter passenger service, and the costs incurred, obviously had an adverse effect on Reno. Within five years he sold his escalator patents to Otis and returned to the United States of America in 1911.*

The installation engineers, employed by The Reno Electric Stairways and Conveyor Limited, were Messrs. G. Aston and Son, Engineers, of Eagle Wharf Road, London. The patent for the drive chain design of this machine was held by William Henry Aston and it seems that the moving walkway could have been an amalgamation of Reno and Aston's inventions.

Whatever the initial company organisational structure had been, its impact had aroused interest within the home market. In 1903, the firm of Waygood and Otis Limited bought a one third stake in the Reno Company.

Reno and Aston must have learned a lot from the Earls Court machine and, by 1906 Aston had revised his design for chains to enable them to follow a path which is not straight. This new, improved design, enhanced by a Reno type tracking and pallet arrangement, was used on the Holloway Road Spiral Walkway. This upgraded passenger conveyor was to be installed at Holloway Road Station on the Great Northern, Piccadilly and Brompton Railway. The installation contract was once again awarded to London engineers, Messrs. G. Aston and Son.

The Spiral passenger conveyor consisted of a continuous, moving platform travelling at 100 feet per minute in both the Up and Down direction. The passengers boarded at either the upper or lower landing by stepping directly onto the moving pallets. The moving walkway, ran very much like a conventional escalator, between a fixed balustrade. The inner handrail was fixed but the passenger could hold the outer moving handrail fitted to a metal framed banister.

*Above: Step detail showing rubber treads, set in aluminum ribs, screwed to wooden palettes.*

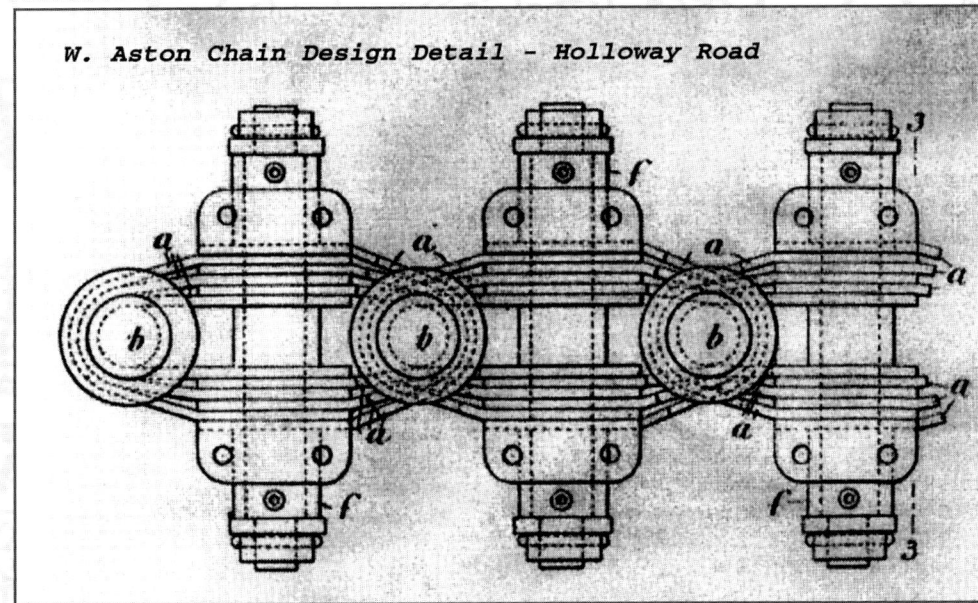

**W. Aston Chain Design Detail – Holloway Road**

The Spiral Walkway was, in fact, like a conventional spiral staircase, except that the outer handrail moved in synchronisation with the walkway. The machine was installed in a 23 foot diameter shaft with a vertical rise of about 35 feet. The passenger walkway was continuous and ran in a clockwise direction, with an inner ascending and an outer descending spiral.

The walkway was guided by wheels fixed to the individual pallets, contained within an angled steel tracking system. The pallets are joined by means of a single handcrafted steel chain, with both vertical and horizontal pin joints.

The outer side links had flanges turned outward to carry the teak slats and cleats that formed the surface of the walkway. The chain was driven by a double steel sprocket wheel, engaging with hardened steel rollers fitted to the ends of the horizontal chain pins. The double sprockets were driven via a system of bevel gears and auxiliary shafts from a main drive shaft.

The handrail was driven in a similar manner, via sprockets driven from the pallet chain, and was simply a chain with an outer covering of India Rubber. It was shaped like an ordinary handrail but the chain had projections on the outer links that ran in a continuous steel guide. The motor and main drive mechanism were located below the lower landing and, in the event of failure, it was pointed out that passengers would still be able to walk down. The speed of one hundred feet per minute meant that the journey time for descending passengers was forty seven seconds and for ascending passengers, forty three seconds.

The drive mechanism however, appears to have been very inefficient. Any differential movement in the main pallet chain may have resulted in the mechanism locking up. Starting the machine would have caused particular problems and may have resulted in it kicking into action in a series of violent jerks.

Although some railway histories have indicated that the machine entered service immediately, the new line opened on the 15th December 1906, considerable doubt must be attached to such statements, as photographs taken before the opening of the line show that, the construction was far from complete. The handrail drive and wire mesh upper balustrade may not have been finished. The Board of Trade Railway Inspector, Col. H.A. Yorke, who inspected the line shortly before opening, commented that the machine was not ready and would need further inspection.

The company does not appear to have requested further inspection; however, it was recorded that a New York consultant, Barclay Parsons, was interviewed to discuss safety issues regarding passenger transportation and the further use of the machine. It appears that the Railway Company may never have entered into a contract for the machine or its installation, it may have only been for demonstration purposes and the project could possibly have been abandoned. There is, in fact, no real evidence that the Reno Spiral Escalator ever entered passenger service, it was unfortunately dismantled in 1911.

Fortunately for posterity, when the machine was dismantled, the lower ring of tracks and support frames were buried intact when the floor level of the lift shaft was raised. It was rediscovered eighty three years later during a station modernisation. The builders who discovered the remains did not really understand the significance of the find and, although, it was verified as Reno's spiral by Mike Horne and Paul Hadley, it was not properly dismantled or surveyed.

**J.W. Reno**
*Configuration of track and handrail drive for the Spiral Walkway.*

999,885.

Patented Aug. 8, 1911.
11 SHEETS—SHEET 2.

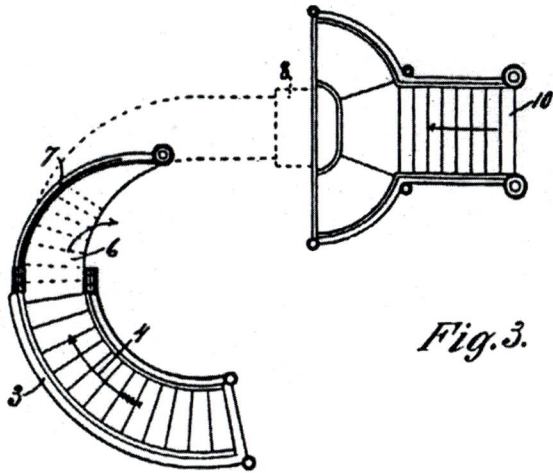

Fig. 3.

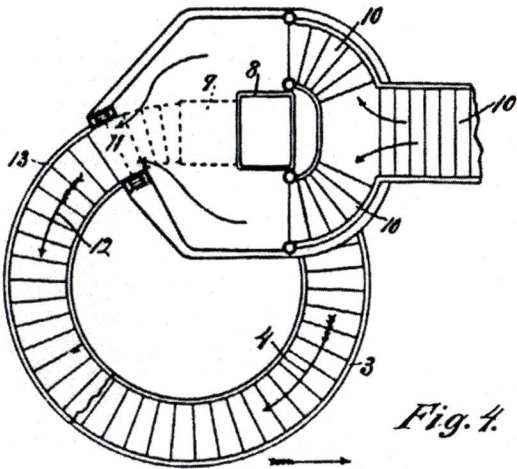

Fig. 4.

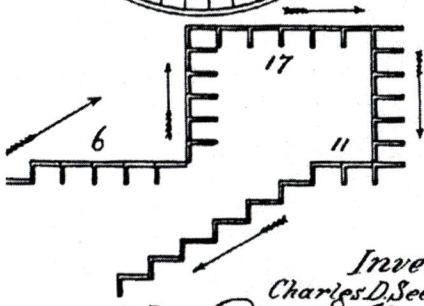

Inventor:
Charles D. Seeberger
By Coburn McRobert
Att'ys.

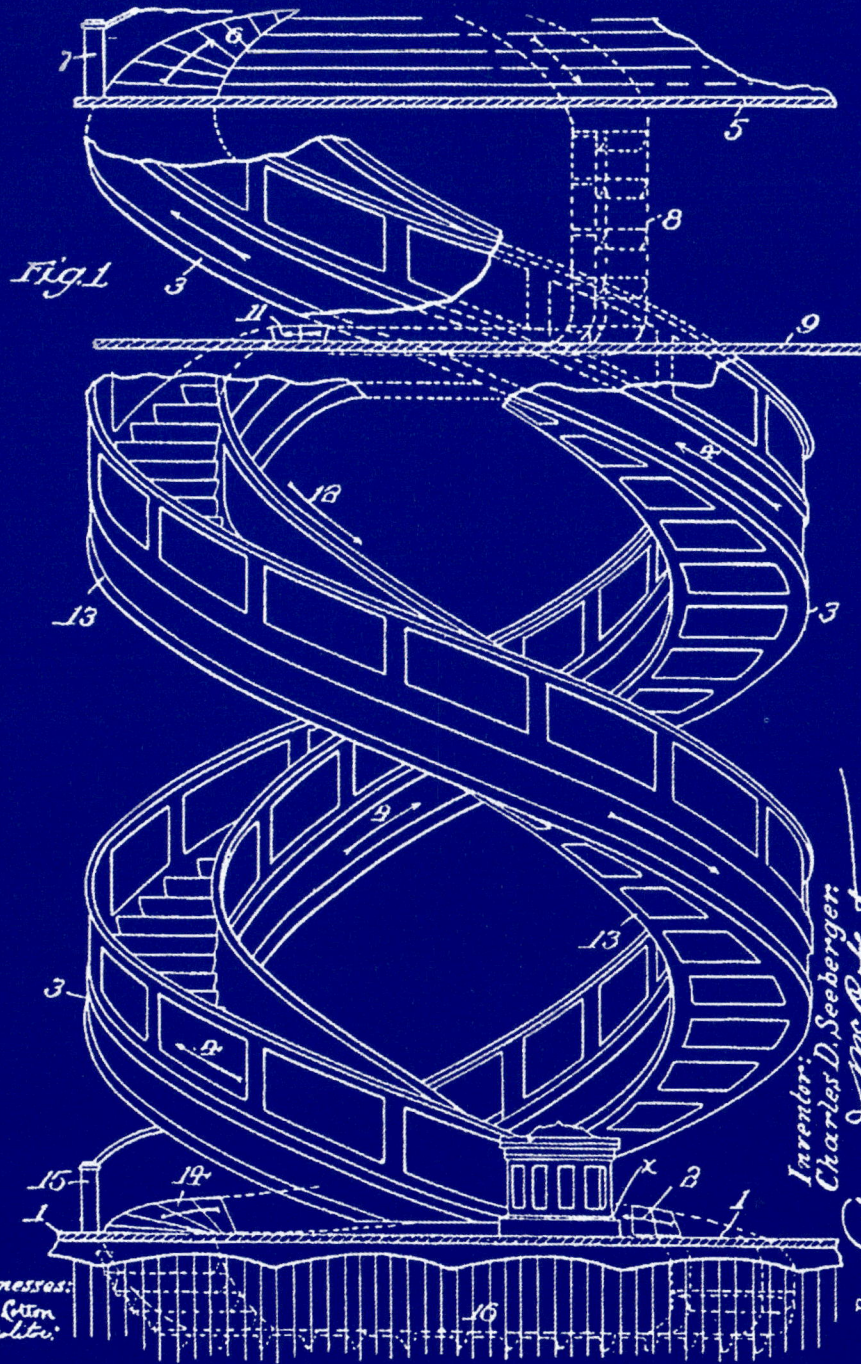

Fig. 1.

17584

Patented Aug. 8, 1911.

999,885.

Inventor:
Charles D. Seeberger
By Coburn McRobert
Att'ys.

Witnesses:
W. H. Cotton
E. Molter.

25

At the same time as Reno was installing his machine, Seeberger was developing his own Spiral Escalator and was particularly busy during the years 1900 to 1910, when he designed many of the component parts, including the 'C' shaped handrail.

From the turn of the century, Otis was always keen to spot innovative engineering talent and very quickly bought into Jesse Reno's Electric Stairways and Conveyor Limited. They also encouraged Charles D. Seeberger and George Wheeler, the other prominent engineers of the day, to develop their respective designs of passenger conveyors. This encouragement led to a healthy competition between these engineering giants and when the time was ripe, Otis amalgamated their talents and developed the inclined passenger conveyor as we know it today. Had these spiral designs been accepted by London Underground, the experience of travelling down from the street to the platform, would be very different today.

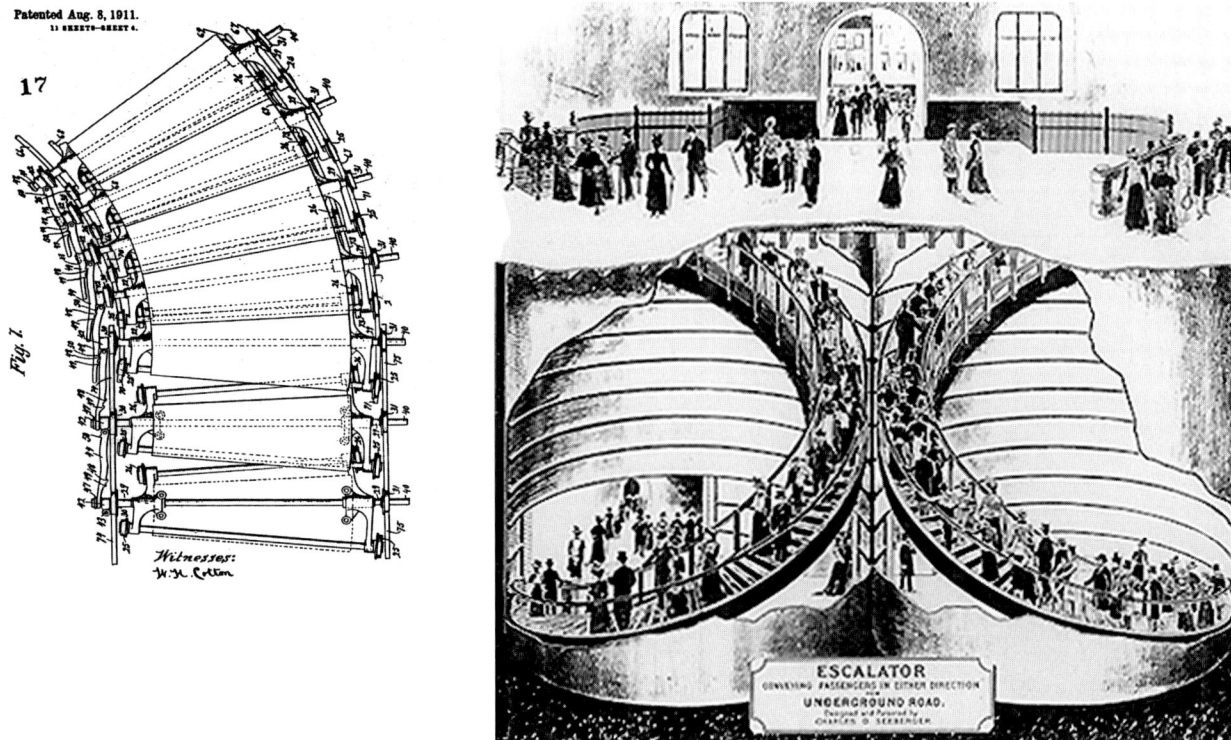

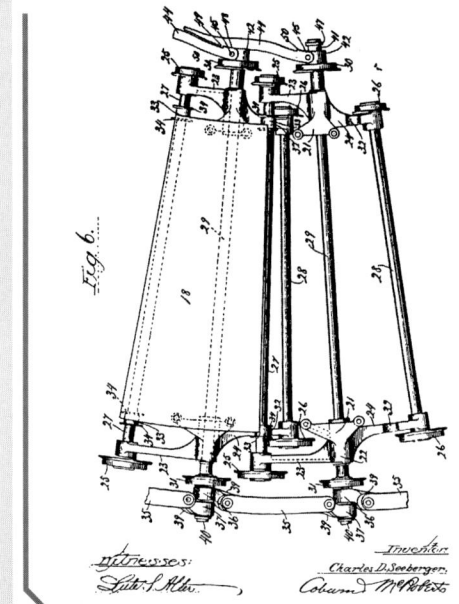

*Left: Seeberger Design Concept drawing for the Underground, with patent details shown either side.*

We will never know if London Underground were ever presented with the superb engineering drawings reproduced on the previous page and shown above, or if they were ever approached to build a Seeberger Spiral. Unfortunately, the lesson learned with the Jesse Reno spiral at Holloway Road had made the Underground engineers sceptical.

# THE SEEBERGER MACHINES 1911-1915

No other escalators were installed on the London Underground until 1911, when Otis installed two Seeberger machines at Earls Court Station. The man in charge of the installation was the highly respected engineer, Mr. J.C Martin, who at that time was an authority on tunnelling. Mr. Martin's retirement was reported in *The Times* on the 27th August 1947, where the journalist waxes lyrical in his praise of the escalator. He describes it thus, 'The moving staircase is not merely an ineffable blessing to weary legs but possesses the fascination of magic in the highest degree.'

He then concludes the article with this magnificent description. 'The escalator is fully as blessed an institution as perpetual motion. Nay, save in the rare case of a technical hitch, it is perpetual motion itself, and should be treated with reverence and gratitude.'

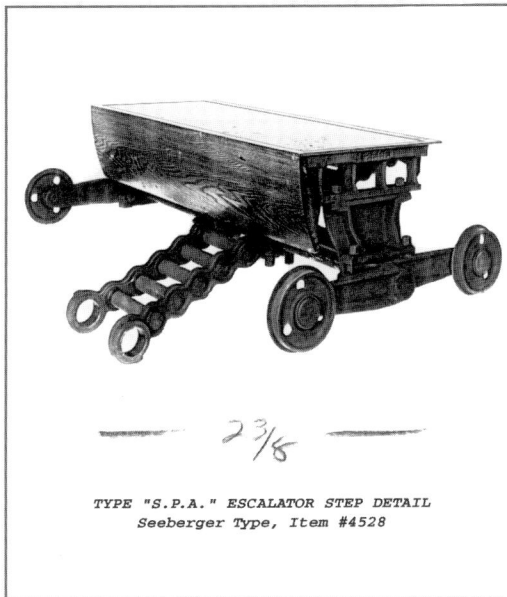

TYPE "S.P.A." ESCALATOR STEP DETAIL
Seeberger Type, Item #4528

*Inventor:*
*Charles D. Seeberger.*
*Coburn & McRoberts*
*attys.*

*Earls Court Station, type 'A' escalator with single step chain.*

The travelling public, at the time of the original installation, would most certainly agree with him. The machines became a tourist attraction with people going to the station just to ride on the escalators. *The Star*, a London newspaper, reported that the escalators at that time were the greatest attraction in London.

The two escalators were prefixed A and were an amalgamation of the Seeberger and Wheeler designs both now owned by Otis. They had flat steps and shunt landings and ran at a speed of 90 feet per minute. The shunt landings required passengers to carry out a side stepping movement when boarding or leaving the escalator. Each step was connected to a single chain running down the centre of the machine.

The chain was kept in tension by a system of counterweights, pulleys and spring loaded tension rods and driven by a large bronze rimmed sprocket situated on the upper main drive shaft. The power being supplied via one of two six hundred volt DC motors directly coupled to a duplex worm-gear. In the event of motor failure, transmission could switch to the secondary drive, enabling continuous, maintenance-free service.

The trusswork was reinforced with sheet steel, virtually enclosing each machine in a steel tank. This led to them being known as 'Tank' type escalators, a description handed down and still used today. The shaft that the first two machines were installed in was inclined to the horizontal at an angle of 26 degrees 23 minutes 16.5 seconds, and all the subsequent machines were built and installed at this angle. This peculiar angle and the massive sheet steel tank, enclosing the escalator, still causes the maintenance engineers immense problems.

The steel tanks are often surrounded by concrete and have virtually become part of the station support steel work. They are still perfectly intact and structurally sound and, although massively over engineered, can become distorted by subsidence. This creates problems when undertaking a major overhaul of the present-day escalators, now built and housed within the old trussworks.

Recently, at Oxford Circus Station, No. 1 escalator was removed from service for nearly one year to enable realignment of the trusswork. The track survey had shown that the upper section of tank was completely deformed and would have to be cut out and completely realigned. Replacement was out of the question and would have meant a station closure which at Oxford Circus was unthinkable.

Marcus Hoffmann de Visme completely re-engineered the upper steel-work, including the floortray supports. This meant the removal of tons of concrete and steel work for replacement by new handcrafted components.

All of the original 'A' types have undergone at least two replacements. The 22 tanks now house a combination of LHDM, HD or MYA escalators, all standard 30 degree machines converted to run in the 26 degree shafts. Of the original machines, one survived until 1953 at Broad Street Station, a remarkable length of service.

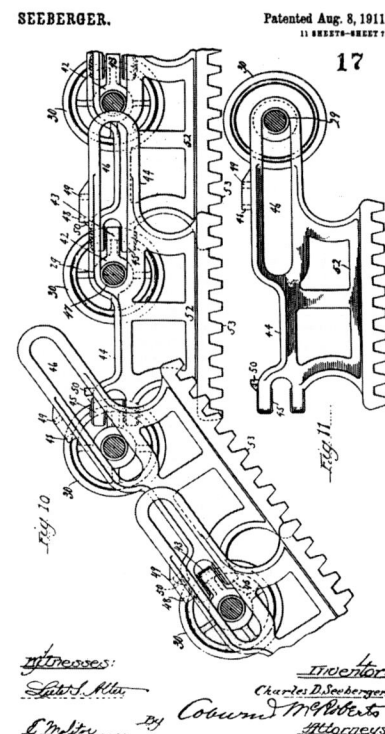

## BUMPER HARRIS

Much has been written about Bumper Harris, the one legged man who is rumoured to have been employed to ride the first escalators and thus prove to the sceptical public that, if a man with one leg can do it, so can you!

This story is not quite true "Bumper" did indeed ride the escalator, but only for the first day. Bumper Harris was the clerk of works for the escalator installation and was asked to ride the escalator as he had an artificial leg (it was not wooden).

*Above: Lowering one of the Tubbing Rings which form the escalator tunnel.*

*Right: Members of the public are invited to test a new design of escalator fabricated at the factory.*

| 1914 | Panama Canal opens. • First electric traffic light in the USA. • Beginning of First World War. |
| 1917 | Bolsheviks storm Winter Palace. • Detergent, the artificial soap invented. • First Pulitzer prizes. |
| 1921-2 | Valentino in The Sheikh. • German Karl Fritsch designs the first autobahn. • First water-skis. |
| 1926 | John Logie Baird displays his television system. • Garnet & Frieda Carter invent miniature golf. |
| 1928 | Fleming discovers penicillin. • Walt Disney creates Mickey Mouse. • Transatlantic telephone. |
| 1931 | Empire State Building completed, NY. • Ernst Ruska builds first electron microscope in Berlin. |
| 1933 | First Boeing 747. • Hitler German Chancellor. • American Charles Darrow invents Monopoly. |
| 1937 | Frank Whittle tests the jet engine. • War between China & Japan. • Nescafe instant coffee. |
| 1939 | Germany invades Poland, Second World War. • Fluorescent lighting. • Insecticide DDT patent. |
| 1940-1 | Battle of Britain and The London Blitz. • Orson Welles, Citizen Kane. • The first aerosol can. |
| 1945-6 | Germany surrenders to Allies. • The ballpoint pen is launched. • United Nations established. |

**Events List**

*As London Underground matured, memorable events continued to shape history as can be seen from the adjoining list.*

# TIME FOR CHANGE 1924 - 1929
## THE M SERIES ESCALATORS
## THE WAR 1939 - 1945

With the great success of the 'A' type machines, the Underground Railway now decided to replace many of its lifts with escalators, a programme to accomplish this was undertaken between 1924 and 1929. The escalator chosen for this replacement was to be known as the LHD. This machine was an improved version of Jesse Reno's Duplex Cleat escalator, taken over by Otis in 1911.

The LHD was a reversible escalator designed for vertical rises up to 60 feet. It outwardly resembled the 'A' type with its wooden balustrade and shunt landings, but the stepband was driven by two chains. Another interesting development of this machine was that the step chain wheels are an integral part of the step chain assembly.

*Left: Tunnelling work in progress on an escalator incline at Trafalgar Square Station in 1925.*

*Above: Men working at the bottom of a triple escalator tunnel, 1928.*

In December 1924, No. 2 escalator at Clapham Common Station was fitted with cleated steps and combplates, making it possible for passengers to move straight on and off at each landing. The new arrangement was so successful that all shunt landing machines were programmed in for modification to accept cleated steps and combplates.

This refurbishment program took place in the 1930's and incorporated eighteen of the 'A' type machines. This took place between 1935 and 1938, resulting in them being known as MA type escalators. In 1937, the first zinc alloy cleated steps were fitted to Kilburn Park No. 2 escalator. These MA type machines continued in service until 1974 when a programme of replacement, starting at Baker Street, was undertaken. The last MA machines to be replaced were at Liverpool Street in 1988.

*Sixty five LHD machines were installed; and Piccadilly Circus Station, with eleven escalators in service from the 10th December 1928, was the showpiece of its day.*

Reno's combplate arrangement, more convenient and user friendly than the shunt landing machines, now became standard throughout the Underground network. This safety feature was important, given the extra demands placed on the system by the increasing numbers of passengers enjoying the new, fast and efficient train service.

As passenger numbers continued to increase other innovations were incorporated into the rapidly changing design to ensure a swift and safe journey tailor-made for Londoners'. The amalgamation of these ideas was about to be realised with the introduction of the 'M' series escalators.

## THE M SERIES ESCALATORS

In the 1930's a new series of machines known as the 'M' type escalators were commissioned. The MA was the modernised Seeberger, the M, MX, MY and MH were designed for vertical rises ranging from 15 feet at Chancery Lane (M) to 90 feet, Leicester Square (MH). These escalators, like the 'L' type, were inclined at thirty degrees.

Many new design features were incorporated on these machines. The MY escalators at Notting Hill Gate were the first machines to be fitted with aluminium balustrades and No. 2 escalator, the first to be fitted with endless handrails.

[ ]

[ ]

[ ]

[ ]

*Left: A prototype MH escalator on test in America.*

*Above and right: Engineers installing newel ends and fitting handrails at Holborn Station.*

Two MX machines were installed at Moorgate Station and were fitted with an electric motorised handrail drive. This proved to be unsuccessful and the machines were converted back to a standard MY chain drive.

Dash pot controls were introduced to the MY auxiliary brakes and load discriminating relays to the MHs. In 1948, Bethnal Green No. 3 was fitted with a 25 HP AC motor.

The first two MH escalators were installed at Archway Station in 1931. They were manufactured in the USA and served as prototypes. All subsequent MHs were manufactured in Great Britain. These machines ran for up to sixty years giving sterling service and performance.

The MH was designed to run at speeds up to 180 feet per minute, with vertical rises in excess of 90 feet. The fast running speed was found to be counterproductive as it was later discovered that escalator passenger carrying capacity was not directly related or proportional to speed and the fast speed setting of 180 feet per minute was reduced to 145 feet per minute. Like their predecessors, the 'M' series machines were fitted with wooden balustrades and steps; however, the step had forged steel yoke arms to overcome fracturing.

*Far right: Holborn Station showing the MH escalators and uplighters typical of this period.*

The 'M' type machine was the baby of this series and covered stations where the rise was between 15 feet (Chancery Lane) up to 35 feet (Turnpike Lane).

The MY was as successful as the MH and covered the middle ground of vertical rise from 16 feet up to 42 feet. The last MY machines were installed at Notting Hill Gate Station in 1960 and were fitted with aluminium balustrade panels.

Today there are still 34 of the 'M' series machines in service, but unfortunately, it seems that time has caught up with them. The new statutory programme of wooden part replacement and fears for the condition of their cast tracks will probably lead to replacement.

I doubt that anyone who has ever worked on these machines has a bad word to say about them. Their design by Lindquist, Handy and Margles is an exemplary blueprint of a fit for purpose, no frills configuration, with an inherent reliability second to none.

Their manufacture and installation (an essential prerequisite if performance is to be maintained over many years) set the standards for future machinery.

*Left: A busy time collecting tickets during the Rush Hour at Highgate Station, 1931.*

*Above: Tiled graphics clearly indicate the exits.*

Sadly, Jimmy Callow, the last Otis manager who worked on these machines recently died and will be greatly missed by all those in the industry who remember him. Jimmy installed the largest MH escalators ever built in Great Britain in the Jarrow pedestrian Tunnel, just after World War II. These escalators are still in service today and are still in immaculate condition.

## THE WAR 1939 - 1945

Although London was subjected to horrendous bombing during the second world war, life in the city carried on. Many escalator manufacturers turned their hands to the war effort and a number of London Underground staff undertook Civil Defence duties, even escalators played their part.

It was, until recently, thought that two escalators were totally dismantled and designated for war service. These machines, assumed to have been St Pauls No. 6 escalator and Chancery Lane No. 5, were supposedly removed from site, taken to the Government Underground Complex at Corsham in Wiltshire and rebuilt.

*Right: A moments relief from the bombing as East Enders enjoy a cup of tea in the relative safety offered deep underground.*

London suffered serious damage during the bombing raids and blitz during the early stages of the war. The greatest fire damage since the Great Fire of London in 1666 was sustained on Sunday night the 29th December 1940. The first wave of bombers hit the city at 6:00 p.m. Within an hour, one hundred fires had been reported and the city's water main was fractured.

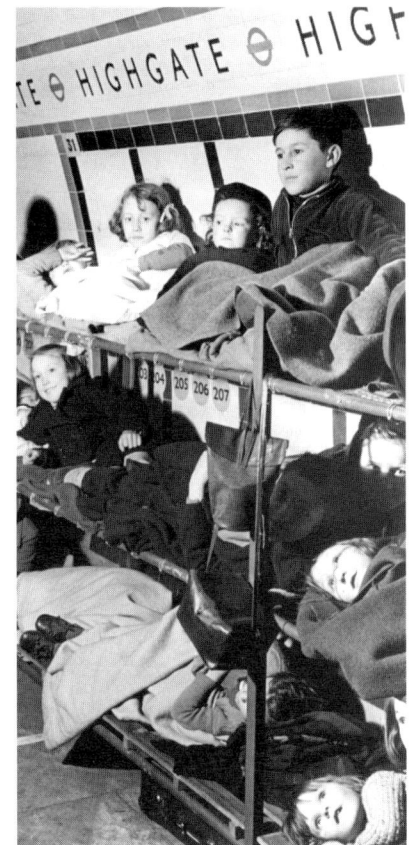

*Left and right: Bank Station destroyed during the Blitz with extensive loss of life and heavy casualties.*

*Above: Children shelter in relative safety sleeping in numbered bunks on Highgate Station platform.*

The fire Brigade had to stand by helplessly as their hoses ran dry. The fire storm raging at 4,000 degrees Fahrenheit, was not confined to the City, warehouses on the south bank of the Thames between London and Tower Bridge burned ferociously. The night sky was lit so intensely that the glow was seen at Bishops Stortford almost thirty miles away. It was 7:30 a.m. over thirteen hours after the raid started, that the fires were contained and it took a further three days of damping down to extinguish them completely.

The Blitz ceased in 1942 for three years but great damage had been done to the City including several of the Underground stations which suffered direct hits. The worst human casualty caused by bomb explosion was at Bank Station. On the 11th January 1941, a bomb went through the station, down a ventilation shaft adjacent to the escalator machine room and exploded at the lower platform.

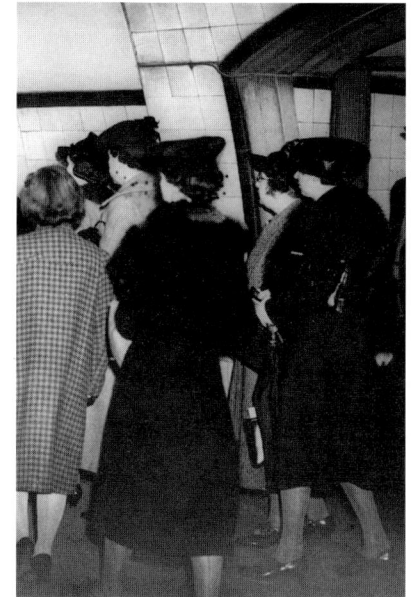

*Left: Sloane Square Station rebuilt after extensive damage caused by a large bomb which fortunately did not damage a culvert carrying the River Westbourne.*
*This would have flooded the system causing even greater loss of life.*

Over one hundred people, sheltering from the air raid, were killed and many more seriously injured, the carnage was terrifying. Many other stations suffered structural damage, Wood Green and Sloane Square being worthy of mention. However, no other air raid incident during the early blitz, in human terms, can compare to the Bank Station bombing.

A terrible civilian accident occurred at Bethnal Green Station, on the evening of the 3rd of March 1943. Local citizens, alerted by the air raid sirens and expecting an imminent air raid, fled into Bethnal Green Station for protection. A young, pregnant woman tripped over on the stairway into the station, and masses of people in their panic to get into the station fell over her. One hundred and seventy three men women and children died in the incident. This appalling civilian disaster was the worst of World War II.

*Above and right: Night after night Londoners shelter in the labyrinth of tunnels which form the Underground system, sleeping on escalators and platforms beneath the the Capital. Piccadilly Circus Station was heavily used, and many survivors still remember the overcrowded conditions associated with these desperate times.*

### Mystery

*Recent research has shown that the escalators installed at Corsham came directly from the manufacturers in 1941, and were presumed to have been those destined for Holborn and St. Pauls. This conflicting amalgamation of information becomes even more compounded when we find that two other machines were also removed from service at about the same time. Holborn No. 4 and one of the MH escalators at Waterloo were definitely not installed as per the original 1930's program and, like Chancery Lane, five were actually installed in 1947.*

43

| | |
|---|---|
| **1947** | Polaroid camera first demonstrated.  •  CIA founded.  •  Williams: A Streetcar Named Desire. |
| **1948** | National Health Service in Britain.  •  Mohandas Gandhi assassinated.  •  Vinyl LP records on sale. |
| **1950-1** | McCarthy witch hunt of communists.  •  Launch of Diners' Club credit card.  •  Colour TV in USA. |
| **1955** | Frozen fish fingers.  •  First nuclear-powered submarine.  •  Lego plastic building blocks invented. |
| **1956** | First transatlantic telephone cable.  •  Elvis Presley's Hound Dog.  •  Ampex launch video recorder. |
| **1957** | USSR launches Sputnik I satellite.  •  Frisbees thrown.  •  Jodrell Bank, first giant radio telescope. |
| **1958** | Aluminium can developed USA.  •  NASA space administration founded.  •  Silicon solar panel. |
| **1959** | Dalai Lama flees Tibet.  •  Cockerell demonstrates the first hovercraft.  •  Antarctic Treaty signed. |
| **1960** | First communications satellite.  •  Hitchcock's film Psycho.  •  Oral contraceptive pill on sale. |
| **1961** | Soviet pilot Yuri Gagarin, world's first spaceman.  •  Oral polio vaccine trials.  •  Berlin Wall built. |
| **1963** | First hypermarket.  •  The Instamatic camera by Kodak.  •  Beatles' first album, Please Please Me. |

**Events List**

*After the World War II it was time for Londoners to rebuild their lives and the capital. London Underground, which played such an important role during the conflict, now had to evolve to meet this new challenge.*

THROUGH NATIONAL SAVINGS →

# POST WAR MODERNISATION
## THE VICTORIA LINE

## Festival of Britain South Bank
### Exhibition 1951

**STEP TREADS**
*Stainless steel riser and treads.*

**HANDRAILS**
*New section guide reduces ware.*

**COMBPLATE**
*Small toothed for smooth running.*

True to the spirit of optimism found at the Festival of Britain, the escalators housed within the Dome of Discovery capable of carrying 8000, people per hour, reflected the futuristic design of the period and were clad in sand-blasted and satin-finished aluminum. The step riser facing plates were polished stainless steel with wooden step treads. These machines were removed after the close of the exhibition and rebuilt at Alperton underground station where they carried passengers from the booking hall up to platform level. This type of escalator was used in large numbers on the Canadian metro system.

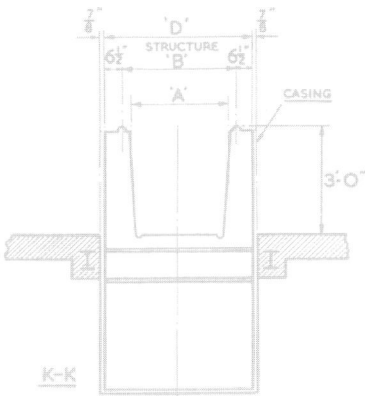

*Above: Alperton Station, Middlesex.*

*Far left: Escalator adapted to fit Alperton Station, a greenfield site.*

*Left: The completed escalator retained its futuristic lines.*

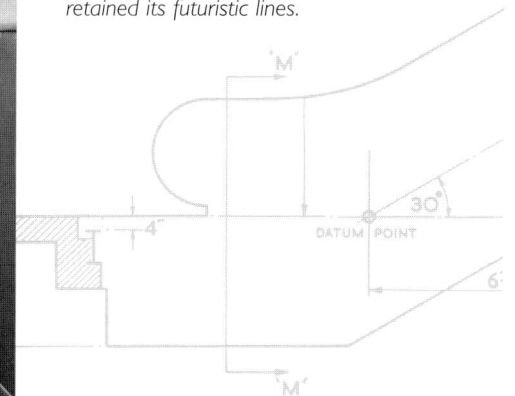

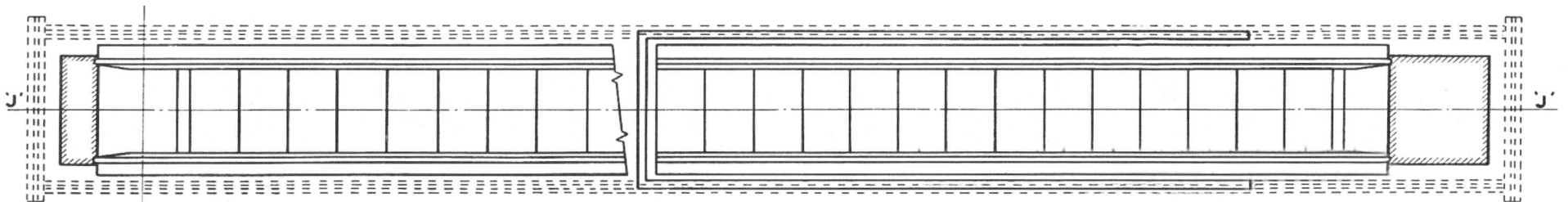

PLAN AT LOWER LEVEL

SHADED PORTIONS AT EITHER END OF ESCALATOR INDICATE REMOVABLE FLOOR COVER PROVIDED BY J.&E. HALL LTD. PURCHASER TO PROVIDE FILLING TO MATCH SURROUNDING FLOOR.

PLAN AT UPPER LEVEL

47

The 1950's provided a period of quiet experimentation and saw the conclusion of the 'M' series escalator installation. Notting Hill Gate was the last station to be fitted with the original design 'M' type machines and the MY escalators installed there were subjected to a degree of experimentation. Aluminium balustrades, endless handrails and sectioned comb plates were all successfully introduced and tested at this site, making the machines a test bed, for the upgraded and modernised machines installed in the sixties.

Also undergoing experimentation were two of the four MX escalators, installed at Moorgate Station. These escalators had been originally fitted with specially designed wooden veneered balustrades, known as Empire Veneer. The balustrade timber was faced with an exotic veneer made from a variety of woods from all over the world, they were elegantly displayed and labelled and gave an aesthetically pleasing finish for the travelling public.

*Left: The modern style in evidence at Waterloo Station. Escalators 4, 5, and 6 serve this busy main line terminal.*

*Above: The main brake mechanism ensures controlled stopping in case of an emergency.*

*Right: Trusswork at Highgate Station will eventually carry passengers to the upper street level, 1957.*

These panels, not only tastefully cultured, but also experimental, were monitored to determine which timber would be most suitable for future escalator balustrades. In 1951, however, they were replaced with plastic laminated panels; the experiment continuing and advancing with the introduction of new materials.

This period is also memorable for the introduction of the first non-Otis manufactured escalator. The escalator was a J&E. Hall machine and set the precedent for other manufacturers to pursue a working relationship with the London Underground.

The "story" of Londons escalators would not be complete without refering to their kindred machines known as Trav-o-lators. The Southern Region of British Railways owned two such machines which operated at Bank and Waterloo & City Railway stations.

After the war a great many people were once again using the London Underground system but the economic position prevented modernisation. In 1955, because of growing congestion, escalator schemes were again considered but found to be too expensive. It was during this period that alternative schemes involving Otis Trav-o-lators were considered.

The design and constructional details of these machines are very similar to those of conventional escalators, but the fundamental difference lies in the fact that, whilst an escalator has a series of inclined steps, a Trav-o-lator has a series of flat platforms which move horizontally.

## THE VICTORIA LINE AND LHD ESCALATORS

The LHD escalator had given exceptional service but was now ready for modernisation. Many of its unique features, such as the step chain wheel being an integral part of the chain, the 'stub' axle step and the pawl brake were retained. In 1963 the escalator was converted to close cleat step and comb arrangement with metal balustrades.

The original LHD machines had a running speed of 110 feet per minute, however, the new updated escalators ran at a speed of 125 feet per minute. A new handrail drive system with endless vulcanised rubber handrails was fitted and the new aluminium balustrading made the machines look ultra modern, a compliment to the 'swinging sixties' reputation London was developing.

*Right: The first LHD-M escalator conversion at the Oval Station.*

The first machine to be converted was Oval No 1 escalator which was returned to passenger service on the 3rd. of September 1963. The programme continued through to the eighties, when the final LHD escalator to be modernised and assume the new prefix LHDM, was at Colliers Wood in 1982.

This nineteen year period saw several significant design changes including the conversion of the cast iron chain wheels to Urethane tyred wheels. The prototype wheel had many teething troubles, several changes were made until the correct design was achieved. The results were worth waiting for as the noise levels and the tracking wear, caused by the cast iron wheels, were reduced.

The Otis staff undertaking the conversion works soon became experts at carefully dismantling the old machines, without disturbing the running of the Station, removing the used component parts and replacing with new. Most of this work was undertaken at night but day shift working could continue if the escalator shafts were correctly hoarded.

Many of the Victoria Line chargehands would swap between working on MHA and MYA machines and the new LHDM conversions. Dennis Butt, Lenny Meredew, Peter Clews and Kevin Smith were all conversant with these machines. The Otis Tester Adjusters Geoff Hutchinson and Dave McGraw (starting out his Otis career) gained extensive experience during this very busy period. The replacement programme concentrated on the central London stations first, gradually covering the outlying districts, finishing up at Colliers Wood on the Northern Line.

*Above: LHD-M escalator, showing the step-chain with tabs and integral wheels.*

*Right: Engineer working on the lower carriage of an escalator at Baker Street.*

At the same time as the LHDM modernisation, London Underground started work on a new Underground extension to be known as the Victoria Line, this extension required 52 new escalators. Otis Elevators were asked to develop two new machines, the MH-A and the MY-A. As their prefixes suggest their construction and design were based on their predecessors, the very successful and reliable MH and MY.

The Victoria Line was the first new Underground Line to be built for many years, running from Victoria to Walthamstow via central London. It was later extended to Brixton via Pimlico and Vauxhall. The Line was opened by Her Majesty The Queen on the 7th March 1963 and was heralded as the most advanced underground railway in the world. The Queen was the first monarch to open an underground railway line since her great grandfather King Edward VII (at the time The Prince of Wales), who opened the first Electric Tube Railway in 1890. He also opened the Central London Railway which later became known as the 'Tuppeny Tube'.

*Lower left: Extensive fire damage inside the escalator machine room at Tottenhan Hale Station, 1968.*

*Below: Staying one step ahead of the traveling public rerquires constant modernisation and refurbishment.*

The new Line caused much disruption in the city as massive modernisation of Stations such as Oxford Circus took place. Many new interesting tunnelling techniques had to be developed, such as those at Tottenham Hale, where the marshy ground prohibited normal 'digging'. The tunnelling machinery could not operate in the waterlogged ground so bore holes were sunk and filled with liquid nitrogen, freezing the ground to enable tunnelling to continue.

Many of the Staff, employed to build the escalators, are still with Otis today. Site Installation Engineers Dennis Butt, Lenny Meredew, Peter Clews and Kevin Smith have well over a hundred years escalator building and maintaining experience between them.

On completion of the Victoria Line, no more escalators were built until 1972, when Otis installed six RAC escalators at South Kensington Station made by Ascinter of France. After this they decided to concentrate production of escalators at their factory at Stadthagen in West Germany.

*Above: The massive 80H.P. electric motor used to drive the MH-A escalator at Seven Sisters Station.*

New machines were to be factory built and then, like their smaller 'store type' cousins, broken down into sections and shipped to England for site assembly. All escalators prior to this were built on site, a type of bespoke engineering, with staged factory inspections of the important component parts prior to site delivery.

The man most closely associated with the Victoria Line works, Ron Button, was in charge of the New Works Inspection. During construction, the Lift and Escalator New Works section, came under the jurisdiction of the Chief Engineer, Lifts and Escalators.

Ron and his Site Inspectors closely supervised all aspects of manufacture, including delivery, installation, commission testing and handover of the Victoria Line escalators. Their contribution to the success of the project was immense.

Now retired, Ron's successful relationship with the contractors and site agents made him one of the most well known and respected members of the escalator world. Any one who met Ron will have their own stories to tell but I will never forget my introduction to him.

It was at Oxford Circus Station and John Davis and I had unearthed a major problem with No2 escalator's upper main drive. To verify our findings, as the machine had only recently been commissioned, the installation Contractors has been notified and we had arranged a site meeting. As John and I waited on site, the escalator machine chamber was suddenly invaded by an excitement of Design and Installation engineers. This array of senior Management turned out to be the escalator manufacturing engineers, led by our own Ron Button.

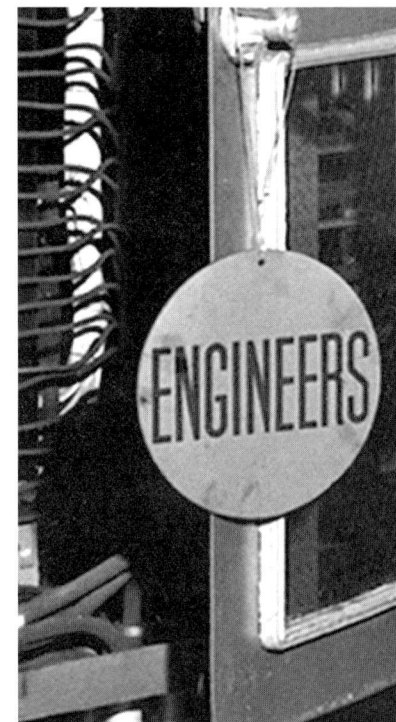

*Left: As part of the ongoing modernisation of the Underground, new escalators are installed at Walthamstow Station.*

*Above: A warning sign is hung on the door of an electrical controll cabinet in the machine room at at Waterloo Station.*

Ron, always the super smart executive, introduced himself and then almost casually climbed into the escalator truss work, (without overalls), and began directing the design experts.

This ability of leading from the front, with confident and inimitable style, I was later to witness many times. Ron could operate equally well in both Directors Board Room or shop floor and this blend of practical experience coupled with site familiarity commanded both respect and acclaim throughout the Lift and Escalator fraternity.

*Above: Ron Button, New Works Engineer, well known and respected, was the man most associated with the Victoria Line works.*

*Right: Lower landing trusswork for escalators Nos. 7 and 8 at Euston Station.*

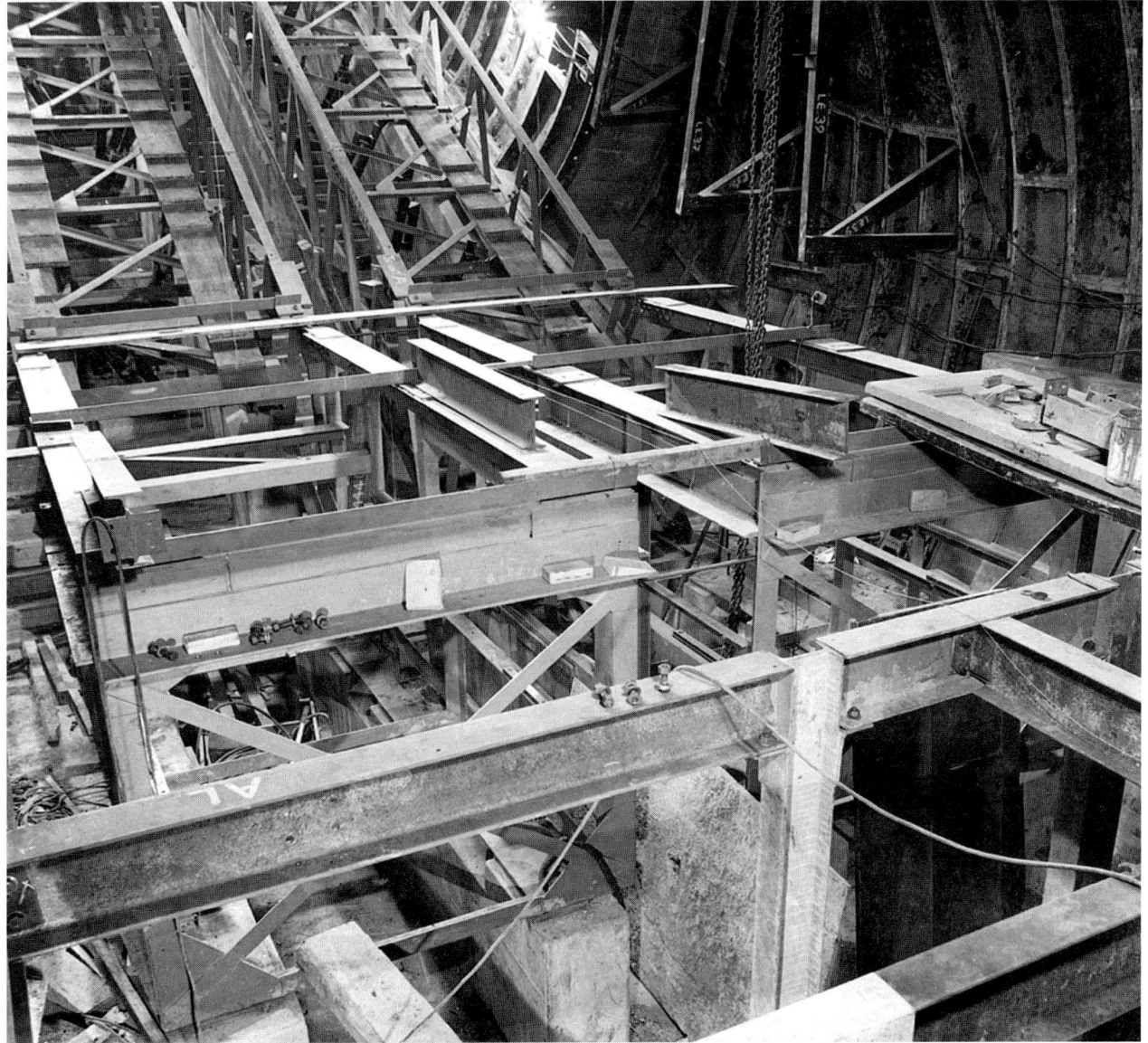

| | |
|---|---|
| **1964** | Japanese bullet train runs. • Beatles' tour USA. • IBM 360 series using wafer thin silicon chips. |
| **1965** | Mary Quant creates miniskirt. • Malcolm X assassinated. • First electronic telephone exchange. |
| **1967** | Israel / Arab states: Six-Day War. • First heart transplant. • ing Constantine of Greece deposed. |
| **1968** | Martin Luther King assassinated. • Kubrick: 2001: A Space Odyssey. • Student unrest, Europe. |
| **1969** | Apollo 11 lands on the moon. • Woodstock festival. • Maiden supersonic flight of Concorde. |
| **1970** | Wide-body 747 jumbo jet. • Global Positioning System using 21 satellites. • Heart pacemaker. |
| **1971** | East Pakistan becomes Bangladesh. • First digital watch. • Soviet space station Salyut 1 in orbit. |
| **1973** | US forces withdraw from South Vietnam. • Aboriginals given vote. • World Trade Center NY. |
| **1974-5** | Nixon resigns over Watergate. • First video game. • US Apollo and Soviet Soyuz spacecraft link. |
| **1977** | First autofocus camera. • Neutron bomb developed. • Bell Telephone system uses optic cables. |
| **1979** | Trivial Pursuit. • The British 125 High Speed Train. • Mobile cellular phone. • The Walkman. |

**Events List**

*With improved living standards, and the cultural revolution of the sixties, public expectation was now very high.*
*As a result London Underground had to respond by improving its service to the standard demanded by Londoners and visitors alike.*

**FACTORY ASSEMBLED ESCALATORS**
MAINTENANCE ENGINEERS

## THE FACTORY ASSEMBLED ESCALATORS

The closure of the British Otis Factory at Liverpool, meant that London Underground would no longer have the luxury of individually crafted machines, but would have to rely on new factory assembled escalators. Although inexperienced with these machines, the Otis installation staff were very quickly undertaking 'in truss' conversions, manoeuvring sections on their side through small station passageways and lowering, complete assemblies into machine chambers. These first German escalators were prefixed HD and were installed by the Otis London Office staff, at Charing Cross Station.

*Left: Building a PH Prototype escalator in close association with London Underground at the Pantin factory.*

*Above: Massive cogs already splined onto their shafts, ready for lifting. Note the scale of the figure in the background.*

The factory assembly technique, incorporating installing the machine in pre-built sections, had also been developed by other manufacturers, which together with competitive tendering, provided opportunities for companies such as Eggers Keherhan to commence installation work on the Underground Railway at Old Street station.

The French company CNIM, installed two machines at Kentish Town Station and Pantin, an Essex conveyor manufacturer, built an escalator designed by London Transport. Pantin was soon bought out by APV Baker, who in turn was soon to find itself incorporated into the massive O&K escalator consortium.

*Above and Right: One of the few factory assembled, heavy duty Eggers Kehrhahn escalators, to be installed in Great Britain, at Old Street Underground Station.*

Competition between respective manufacturers meant more choice and healthy rivalry meant innovating development of engineering practice and expertise. Lighter truss work, ladder step chain, guidance ramps and hydraulic braking systems were all introduced as a result.

This competitive edge, once introduced was to continue through to the nineties. This diversity of design, although not always successful, now complemented by high quality manufacture, meant cost effective and reliable machines were readily available.

## The Otis Elevator PLC

Before the introduction of escalators Otis already built and supplied Station lifts to the underground. They became the first Company to install escalators on London's Underground. The preceding chapters have all concentrated on Otis machines and the history of underground escalators mirrors that of Otis machines up until the 1980's.

The Otis London office, now run by Mike Hirst and Dave McGraw has worked with London Underground since the installation of the first machines at Earls Court. It still plays a major role in the installation and maintenance of many of the Underground's machines. The London Underground branch office, is situated at their Clapham Road headquarters, with Mike Hirst as the London General Manager.

*Above: Otis manager, A.D. Smith photographed during a visit to the Oval Station in September, 1963.*

Mike's team is composed of, Maintenance Manager Dave McGraw, his Projects Manager Richard MacDonald, Technical Manager Mike Harrington and Safety and Quality Manager Duncan Hacon. The whole team has many years of experience and can call upon the expertise of their parent company, United Technology, if required.

To enable them to offer both custom built and factory assembled machines, Otis London teamed up with a Scottish Company, Machine Tool Engineering (MTE) of Blantyre. The Muir brothers, John and Jim, who own M.T.E., worked very closely with Otis engineers to produce a new escalator. It was prefixed the MHB and the first machines were installed at Bounds Green. Leicester Square and Waterloo International followed fast on the heels of the initial installation. The Muir Brothers engineering expertise has since won them work on major maintenance projects, escalator repair and design and also component manufacturing contracts.

*Above: Compared with earlier models, this relay control panel was revolutionary in its day and paved the way for future component miniturisation.*

*Right: An Otis MH-B installed at Leicester Square Station in the mid-seventies, and still in service today.*

Otis currently undertake both installation and maintenance work on the Underground, maintaining machines on the Northern, Victoria, Central and Piccadilly Lines. The breadth of experience within the London Otis workforce means that staff can be interchangeable, working on escalator maintenance or if needed on installation. This luxury stems from the longevity of service by core staff, some of who worked on the initial Victoria Line installation.

People like Clem Devoy, a fitter and great friend with whom I worked thirty years ago and Charlie Corchoran, Don Watson and Lenny Van West, together with many other dedicated engineers were responsible for continued development and efficient running of the London Underground, commonly known as The Tube.

## CNIM

Competing fiercely with Otis, for London Underground contracts, is the French Company CNIM. CNIM are based in France, but like Otis, has offices and factories all around the world, indeed many of London Underground's suppliers are multi national companies. The company headquarters are situated in Marseilles and their origins are in marine engineering for the French navy. The CNIM London Office is now run by General Manager Guillaume Penon and his engineering team, from their London City Office in Scrutton Street.

The first installation undertaken by CNIM was at Kentish Town Station in 1983. The installation was managed by a French engineer, Jeane Raimondo who was aided by British Engineers from Dunlop UK, Inclined Passenger Conveyors Ltd. CNIM went on to install machines at Kings Cross and Angel Stations. The Angel installation was a highly prestigious contract to win and these notable escalators were the longest installed on the London Underground. They were designed to carry travelling passengers through a vertical height of approximately thirty metres.

## Pantin / APV Baker

At the start of the eighties London Underground were keen to introduce new competition to the manufacture and installation of escalators. The introduction of competitive tendering 'opened up' the previously single sourced type option. An Essex conveyor firm Pantin, was approached and together with London Underground engineers, a purpose built heavy duty escalator was designed and built.

This machine was prefixed with the letters PH and was first installed at Manor House Station and commissioned on the 22nd. June 1987. Ian Barber was the Engineer in charge of the installation works and when the company was bought by APV - Baker he continued to work for the new firm. APV-Baker moved the production of PH machines from Essex to their Peterborough headquarters. APV's design team led by Graham Pearce incorporated many of their own design ideas to develop the PSX escalator. This newly amalgamated company very quickly won a series of installation contracts and the PSX was installed at Liverpool Street, Chancery lane, Green Park and Warren Street Stations.

APV's success with the PSX led to designs for lighter construction machinery and escalators that could be suitable for both small vertical rise and high vertical rise application. The new escalators were highly successful and a prestigious contract was won to install PSX machines in Washington DC. Hard on the heels of this inroad into the American market, APV also won two contracts for escalator installation at Kings Cross and Kentish Town, on the London Underground.

*Above: Rarely seen inlaid Underground coloured logo, edged in brass, centrally placed within the floor panel at the top of an escalator.*

APV were now also maintaining and refurbishing escalators on the Underground and starting to expand as a company. The manufacture of their machines was now shared between the Peterborough factory and their North Eastern factory on the outskirts of Jarrow. With further expansion imminent, APV appeared to be in strong position but O&K escalators masterminded a successful take over bid.

Escalator manufacture would now be confined to the O&K Great Britain headquarters in Keighley West Yorkshire. APV staff relocation to Yorkshire was now necessary, obviously not everybody could accommodate the move, (family ties, children's education, etc.). Graham Pierce was one of the casualties and has now returned to his original marine engineering background. Escalator design will be much the poorer for the loss of his skills.

*Above: The street outside the Angel with a new facia and sign.*

*Right: The longest escalators installed on the London Underground system were manafactured and installed by CNIM, and are to be found at the Angel Station.*

## O & K Escalators

Already an experienced and highly regarded manufacturer of 'Store' type escalators, O&K began to win orders for public service escalators. They had worked on major projects at air terminals, railways and modern metro systems, but nothing as complex as London Underground. My first introduction to O&K was in 1990 when Dave Mallett and I were introducing Comprehensive Maintenance Contracts for the Underground escalators. Mark Truelove, O&K's General Installation Manager made enquiries about the tendering process, however he had reservations regarding undertaking work on high rise machinery built and installed by other companies.

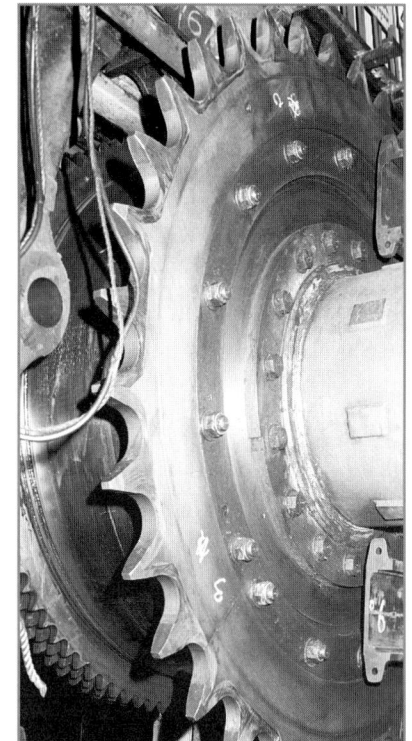

[ ]
[ ]
[ ]
[ ]

We were impressed with his honest approach and endeavoured to offer advice regarding escalator maintenance. O&K did not tender for maintenance but did go on to explore the possibility of escalator manufacture and installation, for future projects.

They soon won an installation contract for the Docklands Light Railway at the Bank Station. Linked to this would be the replacement of some existing high rise escalators on the District and Central lines at Monument and Bank Underground Stations.

*Left: O&K escalators operating at Bank Station.*

*Above: Giant drive sprockets photographed during modernisation after thirty years of hard service at Victoria Station.*

The Bank Station refurbishment project was the first major undertaking on the Underground for O&K, and enabled both designers and engineers to experience the difficulties associated with working on deep level stations in a major city.

In 1993 it was announced that O&K had won the highly prestigious contract for the manufacture and installation of 118 escalators for the Jubilee Line Extension. This was magnificent news for the staff at their Keighley factory and safeguarded production for the next few years with maintenance parts manufacture secured for at least ten years. The Keighley works has many loyal staff with long service records, an indication of the well being and good morale of the company.

*Above and Right: Teams of engineers installing the escalator motor and drive gear at the Bank Station. The photographs give an impression of the difficulties involved when working in the enclosed enviroment of the Underground tunnels.*

O&K are headed by Ron Wanless and Bob Chapman with Geoff Midgley in charge of escalator works and London Underground in particular. Mark Truelove, Mel Lewis and Dave Forest also have tremendous involvement with the London Projects.

Dave Forest is now the senior Design Engineer and has taken over the APV team established by Graham Pierce, his team are based in Keighley, but undertake work throughout the UK. They have been heavily involved on the Jubilee Line installations and commissioning.

65

## THE MAINTENANCE ENGINEERS

Escalator maintenance, like that of all other machinery, is now in its third generation of technical advancement. The first escalators, installed in 1911, were maintained under the philosophy "if it ain't broken, don't fix it'. Due to the sturdiness of design they responded to this philosophy and lasted (Broad St, escalators) for over forty years. The second generation of maintenance was largely developed after the '39 - '45 war and included preventative and planned maintenance. These techniques were also very successful but largely depended on the excellent design of the 'M' series escalators. The third generation, Reliability Centred Maintenance (RCM), based on a system developed by the airlines and refined by John Mowbray, is now being introduced to the Underground.

*Far Left: Mathew Self and John Fryer at Griffith House.*

*Left and Above: Exposed drive sprockets and main bearings allow the massive size and weight of the components to be seen.*

These systems have been undertaken by 'armies' of trained staff since 1911. Escalator maintenance also includes cleaning, lubrication and control of water ingress from the infrastructure. When you consider the area covered by the Underground network you start to understand the colossal amount of work required to keep the system running.

The main depot for lift and escalator maintenance in 1916 was housed in the Baker Street and Waterloo Railway headquarters at the junction of Harewood Place and Lisson Grove. This building later became Marylebone Station but the Lift and escalator department stayed there until 1938 when they moved to Griffith House.

Griffith House, situated at the junction of Chapel Street and Old Marylebone Road, became the Headquarters for London Underground's Lift, Escalator and Pump Division. The building, designed by the Architect A.D. Bolton, was completed in 1929 and was named Lisson House. The name was changed in the 1930's to Griffith House and coincided with the introduction of the first London Transport specific, Lift and Escalator maintenance department. The building (now grade 2 listed) is still fully operational and continues to accommodate the London Underground Lift, Escalator and Pump Manager's maintenance department.

The building provides accommodation and office space for the unit, and a small workshop run by Shift Manager Cliff Garrett. The Pump department, under the guidance of Mick Welsh, has its own workshop where the pump department store and overhaul their emergency equipment.

*Above: The ongoing modernisation and refurbishment of the London Underground system enables passengers to enjoy a clean and efficient service with improved lighting and security.*

*Right: Griffith House completed in 1929 as seen from the Marylebone Road. Now a Grade II listed building, and headquarters of The Lift, Escalator and Pump Division.*

## The Tackle Store Gang

Working for Cliff Garret, the last of the great Shop Foreman, (now known as Shift Managers) are John 'Ferdinand' Davis and Andy 'The Scotch Adder' Hamilton. These very experienced senior Fitters form the backbone of Cliff's empire. Their vast experience 'on the line' ensures that all the staff requiring specialist tools, lifting equipment or hydraulic pumps and jacks are catered for.

Both these fitters have over 25 years service and the wealth of experience that this brings with it. Equipment is never issued without relevant advice accompanying its handover. John Davis may well have built or manufactured the equipment and he never misses an opportunity to elaborate on its construction, use or application. It has even been known for the lads to get John out on site where he invariably takes over and completes the job for them.

John takes all this banter in the manner that it deserves and usually succeeds in teaching the site fitters how to complete a job successfully. This method of practical 'on site' teaching is invaluable to any company and ensures that good working practice is passed on and high morale, in the form of 'toolbox banter' is kept up. These characters and many of their colleagues just like them, are still found over the Lift, Escalator and Pump Managers domain. So long as people of this calibre are employed and nurtured by London Underground, morale, experience and standards will be kept to the highest level and the capital can rest assured that escalator safety and performance will not be compromised.

*Left: Well used tools and equipment fill the Tackle Store ready for any emergency.*

*Above: Bobby Rayner, nicknamed Dirty Bobby, because of his propensity to attract grease and oil.*

Escalator stores and inspection equipment are also housed within the building, ensuring that specialist tools and lifting equipment are always available. LE&PM transport is also based at Griffith House and have an array of specialist vehicles to choose from. The experienced staff are able to move heavy, awkward items of equipment from street level to the required station level.

The Maintenance philosophy for escalators has changed over the years, but so also has the labour force, each generation bringing with it its own characters very much in the 'Bumper' Harris mode. The London workforce reflects the variety of people inhabiting any cosmopolitan major city and it is quite amazing how such a mixture of diverse cultures can combine and produce excellent results. Morale is retained even when coping with seven day weeks and twenty-four hour cover. Most major works undertaken on the Underground take place at night under the most difficult and inhospitable working conditions.

*Above: Andy Hamilton, known as the Scotch Adder, notorious for his ribald quips, funny stories and the infamous 'Indian death lock.'*

*Right: Overhalling the giant drive cogs positioned under the lower landing.*

The Maintenance Department is headed by Mathew Self who together with his team of senior engineers ensure the successful conveyance of passengers from street to platform. Their ceaseless dedication guarantees the often overlooked demand by the travelling public of twenty hours per day running of the machinery.

The 'in house' contractor employs approximately 300 staff whose duties include escalator cleaning, lubrication, service, repair, projects and emergency call out. The hands on staff being supported by technicians, engineers, planners and a 24 hour 'Customer Service Information Centre.'

| | |
|---|---|
| **1980** | Rollerblades are invented. • IBM's personal desktop computer available. • Iran-Iraq war begins. |
| **1981** | AIDS virus identified. • Columbia, first space shuttle. • President Sadat of Egypt assassinated. |
| **1984-5** | First Apple Macintosh computer. • Indira Gandhi killed. • Fuji makes first throwaway camera. |
| **1987** | Disposable contact lenses. • Concerns over fallout from Chernobyl. • Solar powered cars. |
| **1989** | Virtual reality systems available. • Tiananmen Square student protest. • Berlin Wall dismantled. |
| **1990** | Hubble Space Telescope. • S. Africa frees Nelson Mandela. • Genetic engineering developed. |
| **1991** | Allied forces fight Gulf War. • Pollution free electric car. • Volcano, Mount Pinatubo erupts. |
| **1992** | Civil war in former Yugoslavia. • Intelligent Vehicle Highway System, monitors traffic conditions. |
| **1993** | N.Y. World Trade Centre bombed. • Cybershopping on home computer. • Waco Cult siege. |
| **1994-5** | Yitzhak Rabin assassinated over peace moves in Israel. • Channel Tunnel links France and England. |
| **1999** | The first episode of Stars Wars released. • Digital television introduced. • Millennium Dome. |

**Events List**

*As London starts to redevelop after the financial depression, a new optimism takes hold.*
*Many projects are set in motion and London Underground seizes the opportunity to upgrade aging infrastructure and equipment to be ready for the new challenges to come in the 21st Century.*

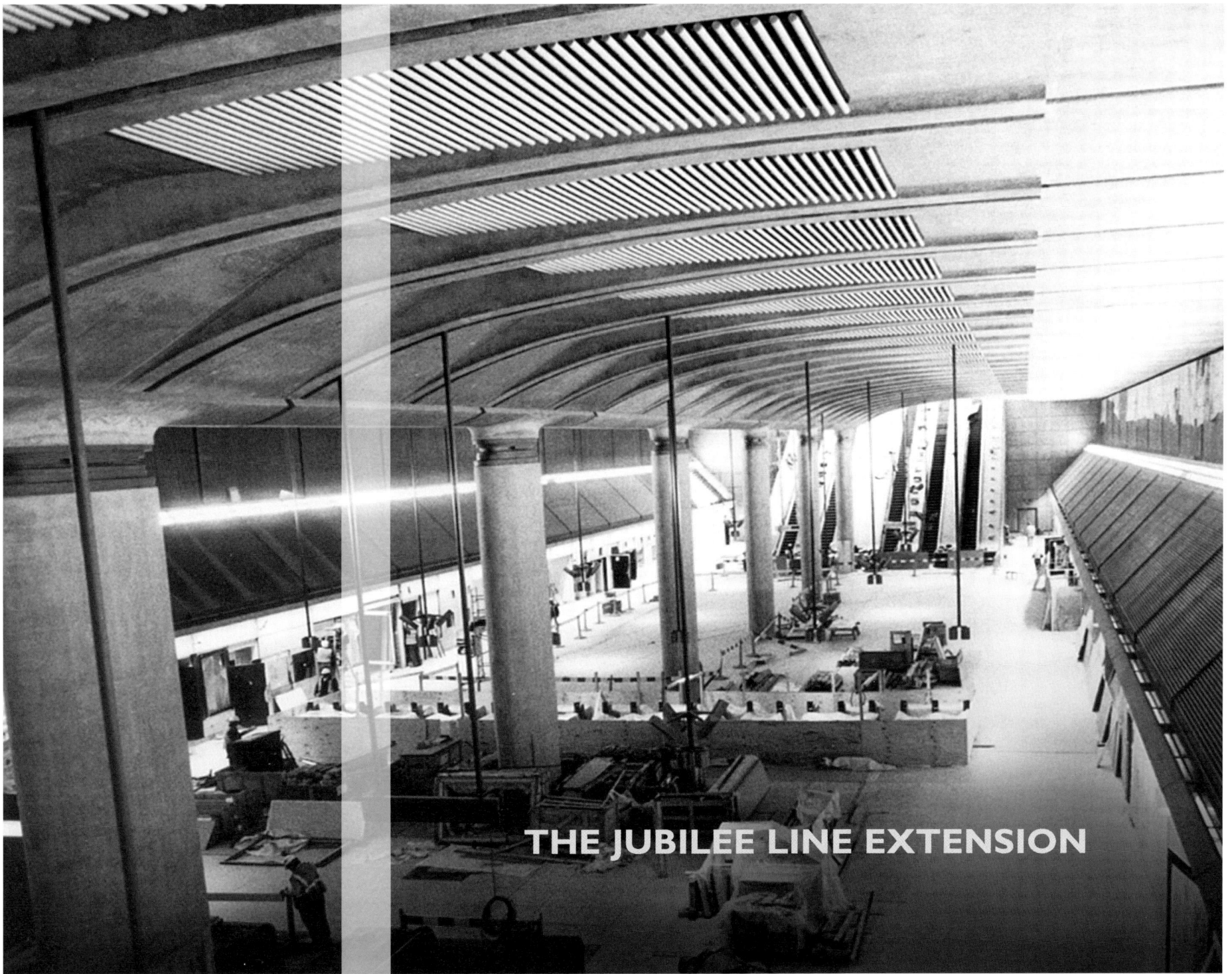

THE JUBILEE LINE EXTENSION

# THE JUBILEE LINE EXTENSION

No major underground tunneling work had been undertaken since the works on the Jubilee Line in the early eighties. The Jubilee Line, originally named the Fleet Line, was opened to commemorate the Queen's Silver Jubilee and was the amalgamation, extension and modernisation of parts of existing tube lines.

In October 1993, the Government gave the go ahead for work to commence on an extension to this Line, that would mean the most significant addition to the Underground Railway in nearly twenty-five years. The Jubilee Line extension would extend the system by a little under ten miles, include eleven stations and cross under the Thames four times. It would join the Jubilee Line at Green Park and go South to Westminster, then East to Greenwich and North under the Thames to Stratford.

*Left: Breaking with tradition, a vast glass envelope allows natural light to penetrate deep underground, illuminating the escalators at Canary Wharf Station.*

*Below: An engineer working at the O&K factory at Keighley dwarfed by one of the new Jubilee Line escalators.*

[ ]
[ ]
[ ]
[ ]

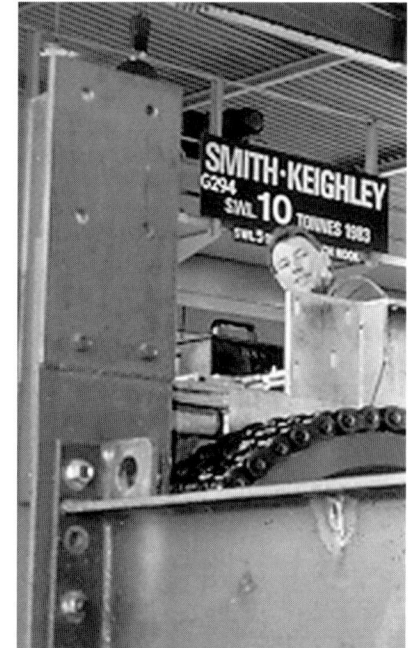

One of the largest construction projects in Europe, it has created a direct link to Docklands and the East End. Now completed, it has brought the Underground Railway to areas of London, like Southwark and Bermondsey for the first time. North Greenwich Station which serves the Millennium Dome, links South East London and Kent with fast easy access to the City and central London. The capital's international links have, as a result, been upgraded at Waterloo and Stratford after extensive referbishment of both stations.

This Jubilee Line Extension, completed in 1999, includes the installation of 118 escalators, 34 lifts and 2 moving walkways. This prestigious installation contract was won by the escalator manufacturer O&K, who will also be responsible for the maintenance. The London Underground team supervising this work and commissioning these machines is led by the Senior Supervising Engineer, Paul Nichols. Paul is more than ably supported by his two hard working and experienced colleagues Kirit Patel and Kevin Seabourne.

*Above: Unusually, these side panels incorporate down lighters to illuminate the treads of this modern looking escalator, and typifies the fresh approach to the design of the Jubilee Line Extension and its equipment.*

*Right: O&K Factory - Keighley.*

Kirit and his assistant engineer Harry Emmet look after the design and manufacturing side of this monumental project. Whilst Kevin, Dermot Tynan, Shahriar Narani, Dave Coster, Alan Groves and John Fitzgibbon oversee the critical phases of site installation and final commissioning of equipment. While this essential work is taking place, Warren Rousseau carries out detailed inspections, quality control and offers advice and support to the team.

As usual, when digging and tunneling under London, many interesting architectural finds are unearthed and the Jubilee Line is no exception. All major tunneling works are rigorously vetted for archaeological discovery and all work must stop, when major finds are discovered so that full and correct archaeological research can be undertaken. When digging the new escalator shaft at London Bridge an early Roman settlement was found, including the remains of three coffins, stacked one on top of the other.

One of the coffins contained the skeletal remains of a woman and child. Under London Bridge viaduct a second century building was found and conserved and a particularly fine miniature Roman 'foot lamp' was found in this same area of Southwark. The Westminster excavation revealed evidence of occupation during prehistoric and medieval times including an Inn, named the Saracen's Head, a gatehouse and a merchant's house in the middle of Parliament Square.

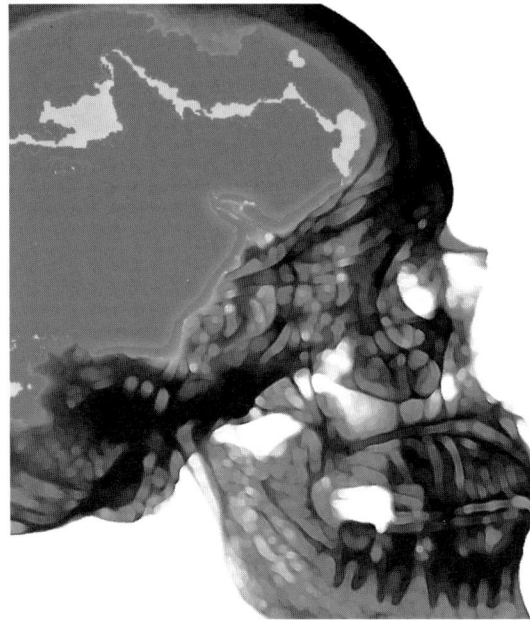

For over a century men have been tunnelling beneath the streets of London to extend the Tube network. Many surprising artifacts have been uncovered, allowing us to glimpse the Capitals historic past.

Tunneling on this scale always causes problems for existing buildings and great care must be taken when the tunnels pass directly below or adjacent to Historic monuments. Existing machinery, like London Bridge Station Northern Line escalators, must be monitored for ground, or existing tunnel movement. This can be fairly unpredictable to gauge, even with today's sophisticated monitoring equipment, so round the clock measurements and movement must be recorded. If certain limits are exceeded then tunneling must stop and the problem rectified before the digging can re-commence.

The 118 escalators supplied by O&K have been designed to meet London Underground's stringent specifications. To ensure good passenger flow in the event of escalator failure or planned maintenance more machines have been used per site.

The Jubilee Line Extension, now completed at the turn of the twentieth century, is a welcome addition to the London Underground railway system and will contribute to the regeneration of Docklands as well as dramatically improving customer choice. It will also relieve congestion on the roads and allow other heavily used Underground lines to function more effectivly by reducing passenger numbers. This massive expansion intriguingly reflects the growth that Charles Tyson Yerkes, oversaw at the turn of the nineteenth centuary. With the new stations and people moving equipement Londoners and visitors to Englands capital city can rest assured that the Tube is in the capable hands of a new breed of escalator and maintenance engineers.

*Above: Installing one of the new factory built escalators at Southwark Station.*

*Right: As the financial heart of the City grew to include Docklands in the East, a more efficient transport system was essential for people living and working in the area. London Underground met this challenge by building the new Jubilee Line extension.*

Just as the magnificence of St. Pauls Cathedral is complimented by the modern city offices and shops of Cheapside, and in the same way as the meritorious architecture of Canary Wharf rubs shoulders with the old dockyards and renovated warehouses of the East End, so the Jubilee Line extension, with its ultra-modern architecture and machinery, sits in perfect harmony beside an older relative; the century old London Underground Railway.

This pleasing blend, so often reflected in the diverse architectural styles of the capital, extends from beneath the streets to the population of this historic city. The excitement felt by the travelling public, initially experienced at the introduction of the first escalators, endures with the continued expansion of the worlds first Underground Railway.

The London tradition of Moving People, like the perceived perpetual motion of its first escalators, goes on forever, with traveller and machine adding to the life and vibrancy of the capital.

*This page: A new beacon burns brightly at Canary Wharf while deep underground work continues on the ticket hall and platform.*

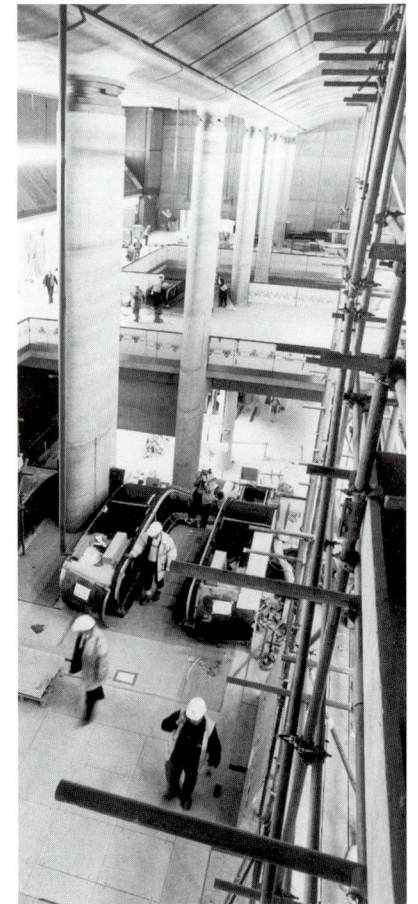

It appears that every fifty years since the Crystal Palace exhibition of 1851, London has hosted major exhibitions to celebrate human achievement.

Attracting visitors from all over the World to such an event places a massive strain on public transport.

The Millennium Dome follows this tradition by providing the arena for what some consider the most important celebration of the age.

London Underground and its engineers have always helped to ensure, through the use of ingenious and innovative people moving equipment, the efficient transportation of passengers throughout the city.

Books that explore the diverse, dramatic and often bizarre history of The London Underground.